I0467990

NOAA Technical Report OAR CPO-1

GLOBAL SEA LEVEL RISE SCENARIOS FOR THE UNITED STATES NATIONAL CLIMATE ASSESSMENT

Climate Program Office (CPO)
Silver Spring, MD

Climate Program Office
Silver Spring, MD
December 2012

UNITED STATES
DEPARTMENT OF COMMERCE

Dr. Rebecca Blank
Acting Secretary

NATIONAL OCEANIC AND
ATMOSPHERIC ADMINISTRATION

Dr. Jane Lubchenco
Undersecretary for Oceans and
Atmospheres

Office of Oceanic and
Atmospheric Research

Dr. Robert Dietrick
Assistant Administrator

NOTICE from NOAA

Mention of a commercial company or product does not constitute an endorsement by NOAA/OAR. Use of information from this publication concerning proprietary products or the tests of such products for publicity or advertising purposes is not authorized. Any opinions, findings, and conclusions or recommendations expressed in this material are those of the authors and do not necessarily reflect the views of the National Oceanic and Atmospheric Administration.

Report Team

Authors

Adam Parris, Lead, National Oceanic and Atmospheric Administration
Peter Bromirski, Scripps Institution of Oceanography
Virginia Burkett, United States Geological Survey
Dan Cayan, Scripps Institution of Oceanography and United States Geological Survey
Mary Culver, National Oceanic and Atmospheric Administration
John Hall, Department of Defense
Radley Horton, Columbia University and National Aeronautics and Space Administration
Kevin Knuuti, United States Army Corps of Engineers
Richard Moss, University of Maryland
Jayantha Obeysekera, South Florida Water Management District
Abby Sallenger, United States Geological Survey
Jeremy Weiss, University of Arizona

Contributors

Benjamin Brooks, University of Hawaii (Manoa)
Mark Merrifield, University of Hawaii
Philip Mote, Oregon State University
William Emanuel, Department of Energy

Workshop Participants

Dorothy Koch, Department of Energy
Jo-Ann Leong, Hawaii Institute of Marine Biology and Oregon State University
Tad Pfeffer, University of Colorado
Sandy Lucas, National Oceanic and Atmospheric Administration

Reviewers

James Titus, Environmental Protection Agency
Jonathan Overpeck, University of Arizona
Stephen Gill, National Oceanic and Atmospheric Administration
John Church, CSIRO
Vivien Gornitz, NASA Goddard Institute for Space Science

Suggested Citation:

Parris, A., P. Bromirski, V. Burkett, D. Cayan, M. Culver, J. Hall, R. Horton, K. Knuuti, R. Moss, J. Obeysekera, A. Sallenger, and J. Weiss. 2012. Global Sea Level Rise Scenarios for the US National Climate Assessment. NOAA Tech Memo OAR CPO-1. 37 pp.

Front cover photo provided by the Greater Lafourche Port Commission

Global sea level rise (SLR) has been a persistent trend for decades. It is expected to continue beyond the end of this century, which will cause significant impacts in the United States (US). Over eight million people live in areas at risk to coastal flooding, and many of the nation's assets related to military readiness, energy, commerce, and ecosystems are already located at or near the ocean.

Past trends provide valuable evidence in preparing for future environmental change but, by themselves, are insufficient for assessing the risks associated with an uncertain future. For example, a number of recent studies project an increase in the rate and magnitude of global SLR. The US Congress recognizes the need to consider future trends in the Global Change Research Act (USGCRA), which calls for a National Climate Assessment (NCA) every four years. This report provides a synthesis of the scientific literature on global SLR at the request of a federal advisory committee charged with developing the next NCA. This report also provides a set of four global mean SLR scenarios to describe future conditions for the purpose of assessing potential vulnerabilities and impacts.

A wide range of estimates for future global mean SLR are scattered throughout the scientific literature and other high profile assessments, such as previous reports of the NCA and the Intergovernmental Panel on Climate Change (IPCC). Aside from this report, there is currently no coordinated, interagency effort in the US to identify agreed upon global mean SLR estimates for the purpose of coastal planning, policy, and management. This is an important gap because identifying global mean SLR estimates is a critical step in assessing coastal impacts and vulnerabilities. At present, coastal managers are left to identify global SLR estimates through their own interpretation of the scientific literature or the advice of experts on an ad-hoc basis. Yet, for a great majority of the US coastline, relative sea level (RSL)[1] has been rising over the past 60 years, consistent with the global trend.

[1] Relative sea level – The height of the sea with respect to a specific point on land.

Scenario Planning

Scenarios do not predict future changes, but describe future potential conditions in a manner that supports decision-making under conditions of uncertainty. Scenarios are used to develop and test decisions under a variety of plausible futures. This approach strengthens an organization's ability to recognize, adapt to, and take advantage of changes over time. Using a common set of scenarios across different regions and sectors to frame the range of uncertainties surrounding future environmental conditions is a relatively new and evolving initiative of the NCA. This report provides scenarios to help assessment experts and their stakeholders analyze the vulnerabilities and impacts associated with possible, uncertain futures.

Probabilistic projections of future conditions are another form of scenarios not used in this report because this method remains an area of active research. No widely accepted method is currently available for producing probabilistic projections of sea level rise at actionable scales (i.e. regional and local). Coastal management decisions based solely on a most probable or likely outcome can lead to vulnerable assets resulting from inaction or maladaptation. Given the range of uncertainty in future global SLR, using multiple scenarios encourages experts and decision makers to consider multiple future conditions and to develop multiple response options. Scenario planning offers an opportunity to initiate actions now that may reduce future impacts and vulnerabilities. Thus, specific probabilities or likelihoods are not assigned to individual scenarios in this report, and none of these scenarios should be used in isolation.

Global Mean SLR Scenarios

We have very high confidence (>9 in 10 chance) that global mean sea level will rise at least 0.2 meters (8 inches) and no more than 2.0 meters (6.6 feet) by 2100.

In recent decades, the dominant contributors to global sea level rise have been ocean warming (i.e. thermal expansion) and ice sheet loss. The relative magnitude

Executive Summary

of each of these factors in the future remains highly uncertain. Many previous studies, including the IPCC, assume thermal expansion to be the dominant contributor. However, the National Research Council (NRC) recently reports that advances in satellite measurements indicate ice sheet loss as a greater contribution to global SLR than thermal expansion over the period of 1993 to 2008. Our scenarios are based on four estimates of global SLR by 2100 that reflect different degrees of ocean warming and ice sheet loss (Table ES-1 and Figure ES 1).

Table ES-1. Global SLR Scenarios

Scenario	SLR by 2100 (m)*	SLR by 2100 (ft)*
Highest	2.0	6.6
Intermediate-High	1.2	3.9
Intermediate-Low	0.5	1.6
Lowest	0.2	0.7

* Using mean sea level in 1992 as a starting point.

Key Uncertainties on Global SLR

At this stage, the greatest uncertainty surrounding estimates of future global SLR is the rate and magnitude of ice sheet loss, primarily from Greenland and West Antarctica. Our Highest Scenario of global SLR by 2100 is derived from a combination of estimated ocean warming from the IPCC AR4 global SLR projections and a calculation of the maximum possible glacier and ice sheet loss by the end of the century. The Highest Scenario should be considered in situations where there is little tolerance for risk (e.g. new infrastructure with a long anticipated life cycle such as a power plant).

Our Intermediate-High Scenario is based on an average of the high end of semi-empirical, global SLR projections. Semi-empirical projections utilize statistical relationships between observed global sea level change, including recent ice sheet loss, and air temperature. Our Intermediate-Low Scenario is based on the upper end of IPCC Fourth Assessment Report (AR4) global SLR projections resulting from climate models using the B1 emissions scenario.

The Intermediate-High Scenario allows experts and decision makers to assess risk from limited ice sheet loss. The Intermediate Low Scenario allows experts and decision makers to assess risk primarily from ocean warming.

The Lowest Scenario is based on a linear extrapolation of the historical SLR rate derived from tide gauge records beginning in 1900 (1.7 mm/yr). Global sea level has risen 0.2 meters (8 inches) over this period of record, and we anticipate at least 8 inches by 2100. The rate of global mean SLR derived from satellite altimetry (1992 to 2010) has been substantially higher (3.2 mm/yr), approaching twice the rate of the longer historical record from tide gauges. However, the 18-year altimeter record is insufficient in duration for projecting century-scale global SLR. Trends derived from shorter records are less reliable as projections because they are affected by interannual and decadal climate and oceanographic patterns that are superimposed upon the long-term rise of global sea level. The Lowest Scenario should be considered where there is a great tolerance for risk.

There is a highly significant correlation between observations of global mean SLR and increasing global mean temperature, and the IPCC and more recent studies anticipate that global mean sea level will continue to rise even if warming ceases. Our Intermediate-Low and Lowest Scenarios are optimistic scenarios of future environmental change assuming rates of ice sheet loss and ocean warming slightly higher or similar to recent observations.

RSL is highly variable over time and along different parts of the US coast. Changes in vertical land movement and ocean dynamics may be applied with different degrees of confidence based on available regional or local data. For example, changes in RSL observed over multiples decades in the Pacific Ocean basin demonstrate that regional-to-global-scale climate factors can cause ocean basin-scale patterns that may persist for a few decades. In the Pacific Northwest, upward vertical land movement reduces RSL on the coast of Alaska and parts of Washington and Oregon.

In the Mississippi River Delta, coastal subsidence increases RSL. As such, individual regions should expect a set of additional processes beyond global SLR that may influence estimates of regional and local sea level change. To provide the level of detail required for sound coastal assessments, regional and local experts, including the NCA chapter authors, are needed to evaluate regional and local ocean dynamics and vertical land movement and make specific adjustments to global scenarios.

SLR and Coastal Flooding

It is certain that higher mean sea levels increase the frequency, magnitude, and duration of flooding associated with a given storm, which often have disproportionately high impacts in most coastal regions. Extreme weather events will continue to be the primary driver of the highest water levels. However, a consensus has not yet been reached on how the frequency and magnitude of storms may change in coastal regions of the US. The greatest coastal damage generally occurs when high waves and storm surge occur during high tide. In many locations along the US coast, small increases in sea level over the past few decades already have increased the height of storm surge and wind-waves. Thus, considering the impact of different weather events combined with scenarios of SLR is crucial in developing hazard profiles for emergency planning and vulnerability, impact, and adaptation assessments.

Conclusion

Based on a large body of science, we identify four scenarios of global mean SLR ranging from 0.2 meters (8 inches) to 2.0 meters (6.6 feet) by 2100. These scenarios provide a set of plausible trajectories of global mean SLR for use in assessing vulnerability, impacts, and adaptation strategies. None of these scenarios should be used in isolation, and experts and coastal managers

should factor in locally and regionally specific information on climatic, physical, ecological, and biological processes and on the culture and economy of coastal communities. Scientific observations at the local and regional scale are essential to action, and long-term coastal management actions (e.g. coastal habitat restoration) are sensitive to near-term rates and amounts of SLR. However, global phenomena, such as SLR, also affect decisions at the local scale, especially over longer time horizons. Thousands of structures along the US coast are over fifty years old, including vital storm and waste water systems. Thus, coastal vulnerability, impact, and adaptation assessments require an understanding of the long-term, global, and regional drivers of environmental change.

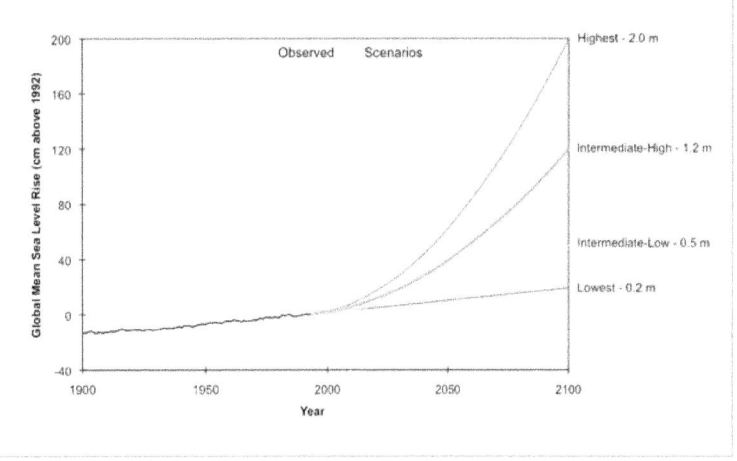

Figure ES 1. Global mean sea level rise scenarios. Present Mean Sea Level (MSL) for the US coasts is determined from the National Tidal Datum Epoch (NTDE) provided by NOAA. The NTDE is calculated using tide gauge observations from 1983 – 2001. Therefore, we use 1992, the mid-point of the NTDE, as a starting point for the projected curves. The Intermediate-High Scenario is an average of the high end of ranges of global mean SLR reported by several studies using semi-empirical approaches. The Intermediate Low Scenario is the global mean SLR projection from the IPCC AR4 at the 95% confidence interval.

Main Report

1. Report Background

Global sea level rise (SLR) has been a persistent trend for decades. It is expected to continue beyond the end of this century, which will cause significant impacts in the United States (US). Over 8 million people live in areas at risk to coastal flooding (Crowell et al. 2010). Along the Atlantic Coast of the US alone, almost 60 percent of the land within a meter of sea level is planned for further development, with inadequate information on the potential rates and amount of SLR (Titus et al. 2009). Many of the nation's assets related to military readiness, energy, commerce, and ecosystems that support resource-dependent economies are already located at or near the ocean, thus making them exposed to SLR.

Evidence for global mean SLR has been increasingly documented in assessments of the Intergovernmental Panel on Climate Change (IPCC), the National Research Council (NRC), and a growing body of peer-reviewed scientific literature (IPCC 2001, IPCC 2007a, Church et al. 2011). The IPCC is currently in the process of revising projections of global mean SLR, drawing on the knowledge and data of the international science community. Similarly, the NRC recently established a panel to assess SLR in California, Oregon, and Washington. State and local governments and the US Army Corps of Engineers (USACE) also are developing or have already developed their own SLR projections for coastal planning, policy, and management from the existing body of scientific literature (USACE 2011, SFBCDC 2011, SFRCCC, 2011, MDEP 2009, MCCC 2008, Horton 2010, Horton 2011).

There is currently no coordinated, interagency effort in the US to identify agreed upon estimates for future global mean SLR for the purpose of coastal planning, policy, and management. This is an important gap because identifying global mean SLR estimates is a critical step in assessing coastal impacts and vulnerabilities. At present, coastal managers are left to identify global SLR estimates through their own interpretation of the scientific literature or the advice of experts on an ad-hoc basis.

For these reasons, the NCA Development and Advisory Committee (NCADAC), a federal advisory committee charged with writing the next NCA, have requested this report. The Scenarios Working Group (SWG) of the NCADAC is considering climate, sea level rise, land use-land cover, and socioeconomic scenarios. This report provides a synthesis of the scientific literature on global SLR and a set of four scenarios of future global SLR. The report includes input from national experts in climate science, physical coastal processes, and coastal management.

2. Using Scenarios for Coastal Risk Analysis

Scenarios do not predict future changes, but describe future potential conditions in a manner that supports decision-making under conditions of uncertainty (Moss et al 2010, Gray 2011, Weeks et al 2011). Scenarios are used to develop and test decisions under a range of plausible futures. This approach strengthens an organization's ability to recognize, adapt to, and take advantage of changes over time. Using a common set of scenarios across different regions and sectors to frame the range of uncertainties surrounding future environmental conditions is a relatively new and evolving initiative of the NCA.

There are many different approaches to developing and applying scenarios (Moss et al 2010, Henrich et al 2010). Scenarios differ in terms of, among other things: varying degrees of qualitative and quantitative information, the goals for use of the scenarios, and the context for the development of the scenario. The purpose may be to explore the consequences of alternative response options; identify impacts of uncertain future conditions; or develop a consensus response option (McKenzie et al 2012). This report provides scenarios to aid assessment experts and their stakeholders in analyzing the vulnerabilities and impacts associated with possible, uncertain futures. Specifically, we provide scenarios to help assess the key uncertainties surrounding estimates of future global SLR: the rates and magnitudes of ocean warming (i.e. ocean warming) and ice sheet loss (NRC 2012, Horton et al 2010).

Probabilistic projections of future conditions are another form of scenarios not used in this report. An early attempt to derive a probability distribution for global SLR combines probabilities associated with major components such as thermal expansion and ice sheet loss (Titus and Narayanan 1995). Given the degree of uncertainty surrounding the different components of future global SLR, this approach is still considered experimental. Horton et al. (2011) provide probable SLR estimates for New York City for the 2020s, 2050s, and 2080s based on certain assumptions. However, this approach requires extensive expertise, computing resources, and substantial prior research, not typically available for many locations of the US. No widely accepted method is currently available for producing probabilistic projections of SLR at actionable scales (i.e. regional and local).

Coastal management decisions based solely on a most probable or likely outcome can lead to vulnerable assets resulting from inaction or maladaptation (Weeks et al 2011, Gray 2011). Given the range of uncertainty in future global SLR, using exploratory scenarios in coastal planning offers an opportunity to overcome decision-making paralysis and initiate actions now that may reduce future impacts and vulnerabilities. Using multiple scenarios encourages experts and decision makers to rehearse multiple future conditions and to develop multiple response options. Thus, we do not assign specific probabilities or likelihoods to individual scenarios because none of these scenarios should be used in isolation. We provide two intermediate scenarios to avoid the interpretation of a single intermediate scenario as most likely. Our approach is consistent with US Army Corps of Engineers (USACE) guidance for coastal decision makers, (USACE 2011) and draws on several commonly used methodologies as the basis for developing scenarios (Henrich et al 2010, Weeks et al 2011).

3. Historical Sea Level Trends

Global SLR is the result of the change in the volume of water in the oceans due primarily to changes in

ocean temperature, melting and increased discharge of land-based ice (i.e. glaciers, ice caps, and ice sheets in Greenland and Antarctica), and changes in runoff (e.g. dam construction or groundwater withdrawal) (Figure 1). Earth's climate, and consequently the sea level, respond to cycles of alternating glacial and interglacial conditions over millions of years, with a periodicity of about 100,000 years (Kawamura et al. 2007).

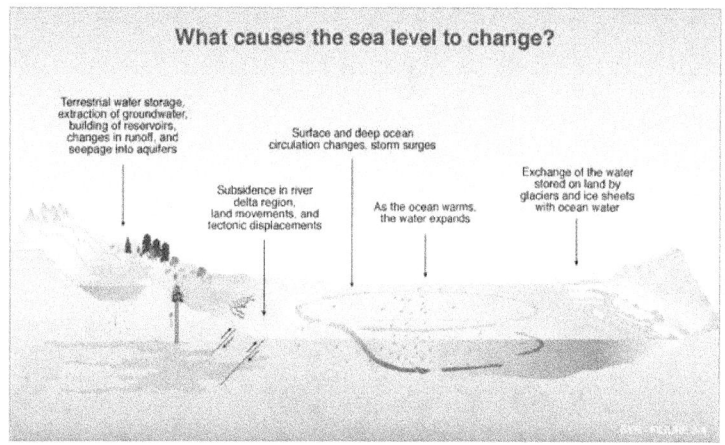

Figure 1. Causes of sea level change. (IPCC 2001)

Geological evidence from approximately 120,000 - 130,000 years ago during the last interglaciation (LIG), a period when the configuration of Earth's ice sheets was similar to that of today, suggests that global sea level rise rates of a meter or more per century are possible (Rohling et al. 2008; Berger 2008). Changes in Earth's orbit during the LIG raised global temperatures to values slightly warmer than present (Otto-Bliesner et al. 2006; Overpeck et al. 2006), causing losses from the Greenland and Antarctic ice sheets that contributed substantially to higher seas (Otto-Bliesner et al. 2006; Overpeck et al. 2006; Hearty et al. 2007; Kopp et al. 2009; McKay et al. 2011). Although sea level variability during the LIG is only a partial analogue for this century, it nonetheless informs the assessment of our scenarios, particularly those that reflect greater ice sheet loss. Geologic evidence of the past 2,000 years suggests that global mean SLR has been relatively stable (approximately -0.1 mm/yr to 0.6 mm/yr) until the late 1800s or early 1900s (Kemp et al. 2011).

Main Report

Since 1900, global mean sea level has been rising at a rate of approximately 1.7 mm/yr as recorded by tide gauges (Church and White 2011). Measurements from satellite altimetry data, beginning in the early 1990s, suggest that this rate has increased to approximately 3.2 mm/yr (Ablain et al. 2009, Church and White 2011). The apparent recent acceleration in global SLR recorded with satellite altimetry may be attributed to a combination of decadal-scale climate and oceanographic patterns, differences in tide gauge and altimetry observation methodologies and measurement distribution, and an acceleration of global SLR associated with global warming.

Satellite measurements reveal important variations in global mean sea level between and within ocean basins (Figure 2). For example, large-scale climate patterns that fluctuate over years to decades, such as the Pacific Decadal Oscillation (PDO), the North Atlantic Oscillation (NAO), and the El Niño Southern Oscillation (ENSO), may cause variations in the Pacific Ocean, the Gulf of Mexico, and the Atlantic Ocean. These variations in mean sea level are referred to as anomalies (see Appendix 1). The cause of these anomalies is still an active area of scientific research.

Tide gauges reveal variations in relative sea level (RSL)[2] resulting from additional local factors such as vertical

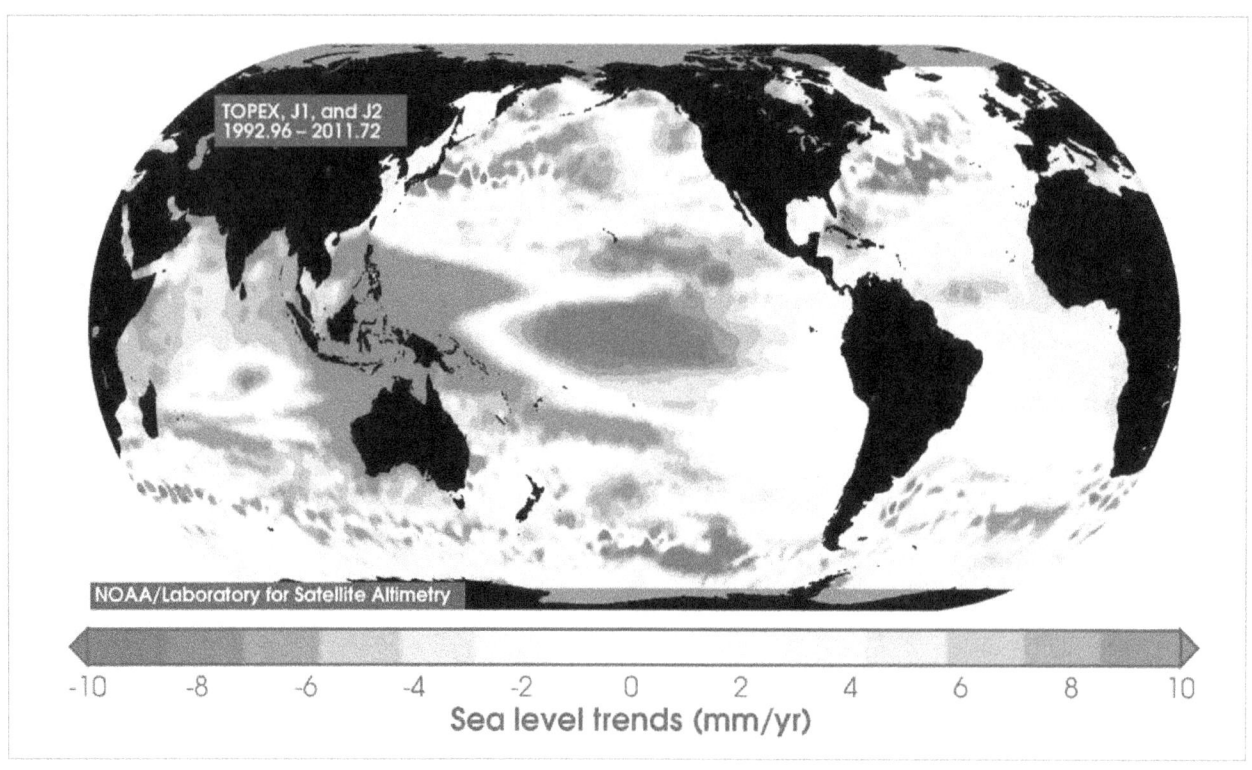

Figure 2. Geographic variability in the rate of global sea level change (1992 to 2010) based on three satellite records: TOPEX, Jason 1 and Jason 2 (Figure source: NOAA Laboratory for Satellite Altimetry – Accessed November 2, 2011).

[2] Relative sea level – The height of the sea with respect to a specific point on land.

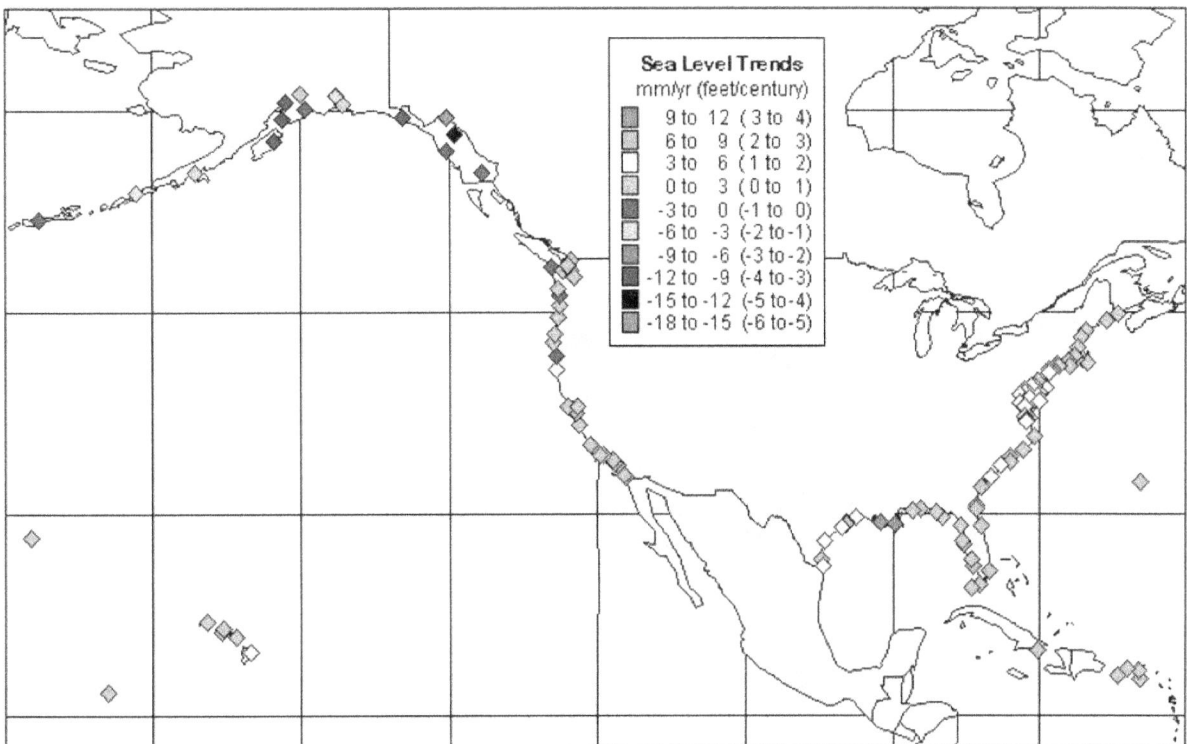

Figure 3. Relative Sea Level (RSL) Variations of the United States (1854 to 2006). Derived from 128 National Water Level Observation Network Stations. Source: Department of Commerce (DOC), National Oceanic and Atmospheric Administration (NOAA), National Ocean Service (NOS), Center for Operational Oceanographic Products and Services (CO-OPS). http://tidesandcurrents.noaa.gov/sltrends/sltrends.shtml (accessed August 16, 2012).

land movement (the uplift or subsidence of the land surface) (Figure 3). For a great majority of the US coastline, RSL has been rising over the past 60 years consistent with the global trend, largely due to limited vertical land movement (VLM) (Figure 4). In the following sections, we discuss patterns and possible causes of changes in RSL and mean sea level anomalies for different ocean basins adjacent to the US coast.

3.1 Regional patterns in the Pacific Ocean Basin

RSL has varied dramatically due to vertical land movement across the eastern North Pacific Ocean (northern California, Oregon, Washington and Alaska) and to decadal climate variability (Figure 3). Both tide gauges and satellite altimetry indicate that RSL has not been rising along the boundary of the eastern North Pacific since the early 1980s (Houston and Dean 2007, Bromirski et al 2011, Merrifield 2011)

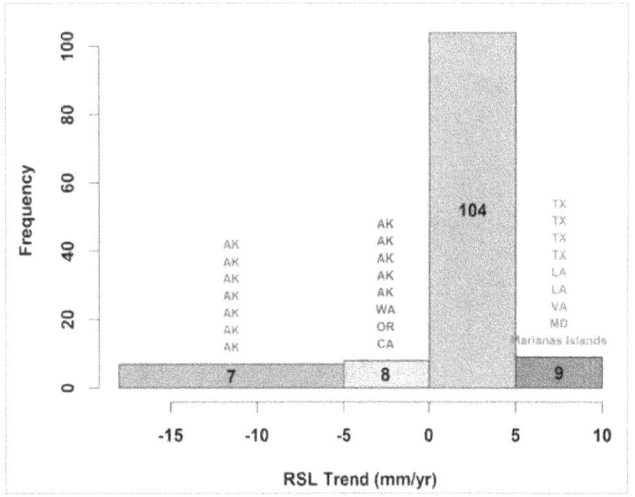

Figure 4. Number of US tide gauges reflecting RSL variations.

(Figure 5a). In contrast, RSL in the western tropical Pacific (Guam, Kwajalein) has been rising much faster than the global trend since the early 1990s (Bromirski et al. 2011, Merrifield 2011). Recent studies suggest that wind patterns may be a primary cause for these variations, though the exact causes are a topic of active investigation.

Bromirski et al. (2011) concluded that RSL trends over the North Pacific are affected by wind stress patterns associated with the PDO, as well as strong, shorter-term fluctuations related to the ENSO (Figure 5b, c). When the PDO shifts, wind stress shifts between the eastern and western boundary of the North Pacific Ocean. RSL rises faster due to greater wind stress curl, the vertical component of wind stress (see Appendix 1) (Bromirski et al. 2011). Bromiski et al. (2011) attribute the recent trend of below global mean SLR along the eastern boundary of the North Pacific to anomalous wind stress curl patterns after the mid-1970s regime shift in the PDO. Recent levels of mean wind stress curl over the entire North Pacific basin have approached pre-regime shift levels, and may signal a return shift in the PDO. Depending on associated changes in wind patterns, this shift could result in a persistent period of RSL rise along the eastern North Pacific that is greater than the global trend (Bromirski et al. 2011).

Similarly, others have concluded that wind patterns have contributed to greater RSL rise in the western tropical Pacific over the past two decades (Figure 5a) (Timmerman et al., 2010; Merrifield, 2011; Merrifield and Maltrud, 2011; Becker et al. 2012). Merrifield (2011) relates this trend to a steady increase in the trade winds since the early 1990s, rather than to interannual ENSO events (Figure 5a). However, Feng et al. (2010) relate this trend to the PDO and low frequency ENSO variability.

The recent shift in the PDO could cause substantial changes in north Pacific winds and, consequently, altered RSL rise. The rate and magnitude of changes in RSL rise would depend on the magnitude of changes in the PDO, the speed and pattern of the trade winds, and other regional anomalies in wind patterns (Bromirski et al. 2011, Merrifield 2011). Sea level anomalies related to ENSO along the eastern Pacific coast can result in as much as 20 cm of RSL rise for an entire winter season (a period of high tides, storms, and large waves) (Bromirski et al 2003, Bromirski and Flick 2008). Regional wind patterns are difficult to forecast over longer time scales (decadal to multi-decadal). Although global MSL rise is a fundamental consideration, RSL height variability along coasts can be more critical and should be taken into account in assessing local impacts, particularly in the North Pacific where the amplitude of interannual variability is high.

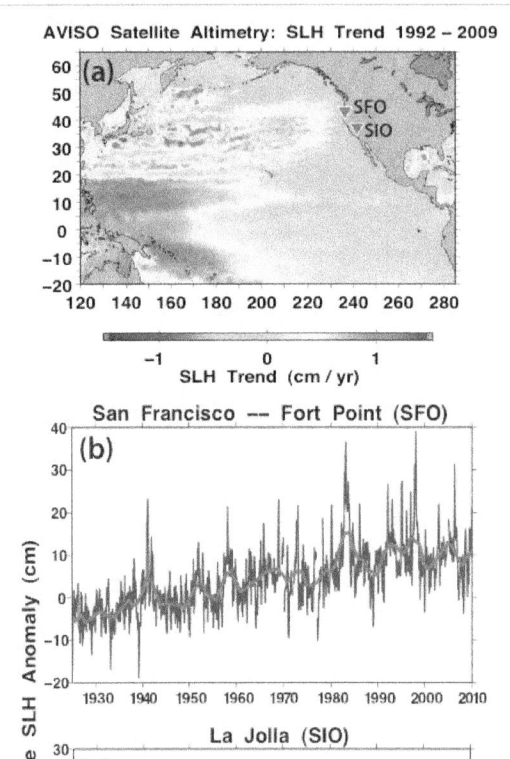

Figure 5. (a) Least squares trends in satellite altimetry sea level height (SLH) across the North Pacific basin over the 1992 to 2009 time period. Observed tide-gauge sea level monthly anomalies, with 3-yr running means (red lines) at (b) San Francisco (SFO) and (c) La Jolla (SIO) (locations in (a)). Note the near-zero trend along the US West coast in (a) that is reflected in the SFO and SIO tide gauge records since about 1980 (Bromirski et al. 2011)

3.2 Regional patterns in the Gulf of Mexico

RSL along the western coast of the Gulf of Mexico (e.g. Texas and Louisiana) has been rising substantially faster (5 to 10 mm/yr) than the global trend (1.7 mm/yr) primarily due to land subsidence. Subsidence rates are highest in the Mississippi River Deltaic Plain where geologic and human-induced factors are compounded. Geological studies suggest that the entire US Gulf Coast region is subsiding as a result of glacio-isostatic adjustment (GIA - see Section 5.1 for further description) (Mitrovica and Milne 2002, Gonzales and Tornquist 2006). Human induced factors include consolidation of thick layers of loose, unconfined sediments; aquifer-system compaction due to fluid withdrawal; and drainage of organic soils.

In addition to high rates of RSL rise, evidence suggests that mean sea level in the Gulf of Mexico has been rising faster than global trend

over the past 60 years (IPCC 2007a). Satellite records suggest that the average elevation of the sea surface in the Gulf of Mexico has increased 3.3 ± 0.4 mm/yr since 1992, with some offshore areas potentially rising as fast as 5.8 mm/yr (Figures 6 and 7). The higher rates of RSL rise across the entire basin may be the result of multi-decadal variability or large basin oceanographic effects given that the Gulf of Mexico is a large, shallow, semi-enclosed basin.

Decadal anomalies in sea level from Galveston, TX to Wilmington, DE are coherent and range ± 5 cm (Figure 8). While 5 cm of sea level may not seem significant, water levels exceeding the height of coastal protection by this amount can cause substantial damage. As seen in the Pacific Ocean region, it is important to consider interannual and decadal anomalies. Experts and decision makers developing risk-averse, worst-case scenarios in specific regions and locations may account for the maximum observed anomaly.

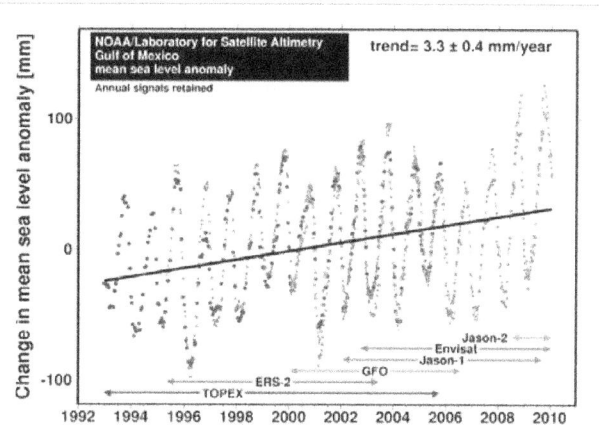

Figure 6. Sea level change in the Gulf of Mexico basin as determined from a combination of six satellite altimetry records (1993 to 2010).

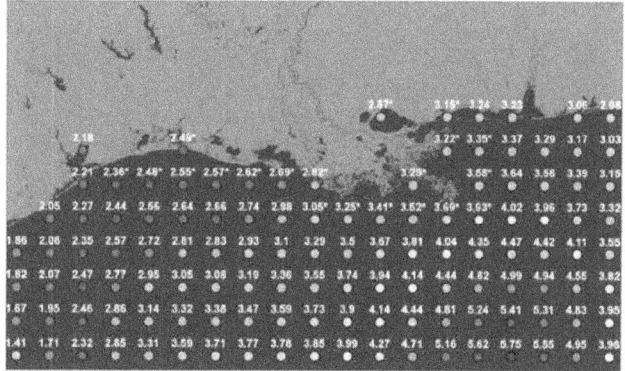

Figure 7. Northcentral Gulf of Mexico satellite altimetry record (1993 to 2011) (Figure source: Brady Couvillion, US Geological Survey, Lafayette, LA).

Main Report

3.3 Regional patterns in the US Atlantic

RSL has been rising along the entire US Atlantic coast, and it has been rising faster in the Mid-Atlantic region (Virgnia, Maryland, and Delaware including the Chesapeake Bay) and the Carolinas primarily due to subsidence (Figure 3) (CCSP 2009). In the Mid-Atlantic region, possible causes of subsidence include sediment consolidation, GIA, groundwater extraction, and tectonics (Poag et al. 2004, Hayden et al. 2008). GIA is less of a factor in the Carolinas, where possible causes of subsidence are groundwater extraction and sediment compaction.

Sallenger et al. (2012) detect a "hotspot" of accelerated sea level rise along the 1,000 km of coast from Cape Hatteras to above Boston and suggest it may be related to circulation changes in the North Atlantic Ocean. Previous analyses along the US Atlantic coast using long records of variable length did not detect this acceleration (Houston and Dean 2011). The presence or absence of accelerations in SLR and the causal mechanisms remain an area of scientific debate. However, the observed rates of RSL rise and the evidence presented by Sallenger et al (2012) and more recently by Boon (2012) are sufficient to suggest that experts and decision makers may consider accelerated rates along the northeastern stretch of coast into their risk-averse, worst-case scenarios.

4. Global Mean Sea Level Rise Scenarios

<u>We have very high confidence (>9 in 10 chance) that global mean sea level will rise at least 0.2 meters (8 inches) and no more than 2.0 meters (6.6 feet) by 2100.</u> Global mean SLR can be estimated from physical evidence (e.g. observations of sea level and land ice variability) (Pfeffer et al. 2008, Katsman et al 2011, Jevrejeva et al 2012), expert judgment (NRC 1987, NRC 2011, NRC 2012), general circulation models (GCMs) (IPCC 2007a, Yin 2012), and from semi-empirical methods that utilize both observations

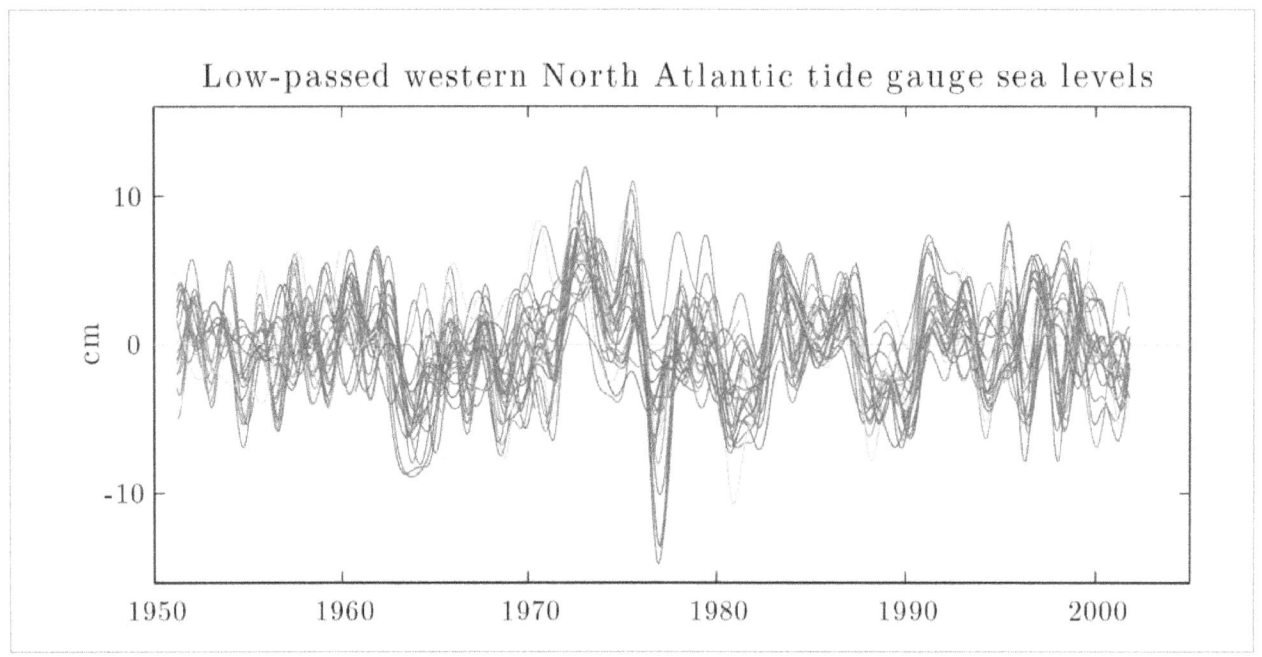

Figure 8. Monthly mean tide gauge sea levels from the Gulf of Mexico (Galveston, TX) to the western North Atlantic (Wilmington, DE). A mean annual cycle and trend are removed, and each series is low-passed with a convolution filter passing >90% amplitude at periods longer than 1.9 years (Thompson 2011).

and GCMs (Grinsted et al. 2009, Jevrejeva et al. 2010, Vermeer and Rahmstorf 2009, Horton et al. 2008, Rahmstorf et al 2012). We base our confidence in the range of estimated global mean SLR on a wide range of the estimates reflected in the scientific literature (Figure 9), using guidance set forth by the NCA (Moss and Yohe 2011, IPCC 2001 and 2007a) (Table 1). This approach has also been applied by the New York City Panel on Climate Change (NPCC 2010).

In recent decades, the dominant contributors to global SLR have been ocean warming and ice sheet loss. Many previous studies, including the IPCC, assume ocean warming to be the dominant contributor. However, the NRC (2012) recently reports that advances in satellite measurements indicate ice sheet loss as a greater contributor to global SLR than ocean warming over the period of 1993 to 2008. Our scenarios are based on four estimates of global SLR by 2100 that reflect different degrees of ocean warming and ice sheet loss (Table 2 and Figure 10).

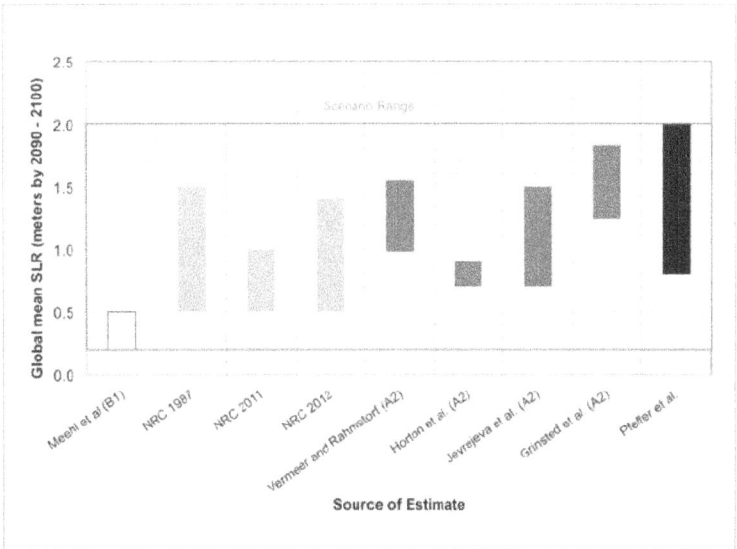

Figure 9. End of century (~2090 – 2100) estimates for global mean sea level rise in meters. Meehl et al (2007) is based on climate model projections for the IPCC and outlined in black. NRC (1987, 2011, and 2012) is based on synthesis of the scientific literature and shown in light gray. Vermeer and Rahmstorf (2009), Horton et al (2008), Jevrejeva et al (2010), Grinsted et al (2009) are based on semi-empirical approaches and shown in dark gray. Pfeffer et al (2008) is a calculation of the maximum possible contribution from ice sheet loss and glacial melting and shown in black.

Table 1. Confidence in the validity of a finding by considering (i) the quality of the evidence and (ii) the level of agreement among experts with relevant knowledge (based on Moss and Yohe 2011).

Confidence Level	Possible Contributing Factors
Very High	Strong evidence (established theory, multiple sources, consistent results, well documented and accepted methods, etc.), high consensus
High	Moderate evidence (several sources, some consistency, methods vary and/or documentation limited, etc.), medium consensus
Medium	Suggestive evidence (a few sources, limited consistency, models incomplete, methods emerging, etc.), competing schools of thought
Low	Inconclusive evidence (limited sources, extrapolations, inconsistent findings, poor documentation and/or methods not tested, etc.), disagreement or lack of opinions among experts

Main Report

4.1 Key Uncertainties on the Global SLR Scenarios

At this stage, the greatest uncertainty surrounding estimates of future global SLR is the rate and magnitude of ice sheet loss, primarily from Greenland and West Antarctica. Our Highest Scenario of global SLR by 2100 is derived from a combination of estimated ocean warming from the IPCC AR4 global SLR projections and a calculation of the maximum possible glacier and ice sheet loss by the end of the century (Pfeffer et al 2008). The Highest Scenario should be considered in situations where there is little tolerance for risk (e.g. new infrastructure with a long anticipated life cycle such as a power plant).

Our Intermediate-High Scenario is based on an average of the high end of semi-empirical, global SLR projections (Grinsted et al. 2009, Jevrejeva et al. 2010, Vermeer and Rahmstorf 2009, Horton et al. 2008). Semi-empirical projections utilize statistical relationships between observed global sea level change, including recent ice sheet loss, and air temperature. Our Intermediate-Low Scenario is based on the upper end of IPCC Fourth Assessment Report (AR4) global SLR projections resulting from climate models using the B1 emissions scenarios. The Intermediate-High Scenario allows experts and decision makers to assess risk from limited ice sheet loss. The Intermediate Low Scenario allows experts and decision makers to assess risk primarily from ocean warming.

Table 2. Global SLR Scenarios

Scenario	SLR by 2100 (m)*	SLR by 2100 (ft)*
Highest	2.0	6.6
Intermediate-High	1.2	3.9
Intermediate-Low	0.5	1.6
Lowest	0.2	0.7

* Using mean sea level in 1992 as a starting point.

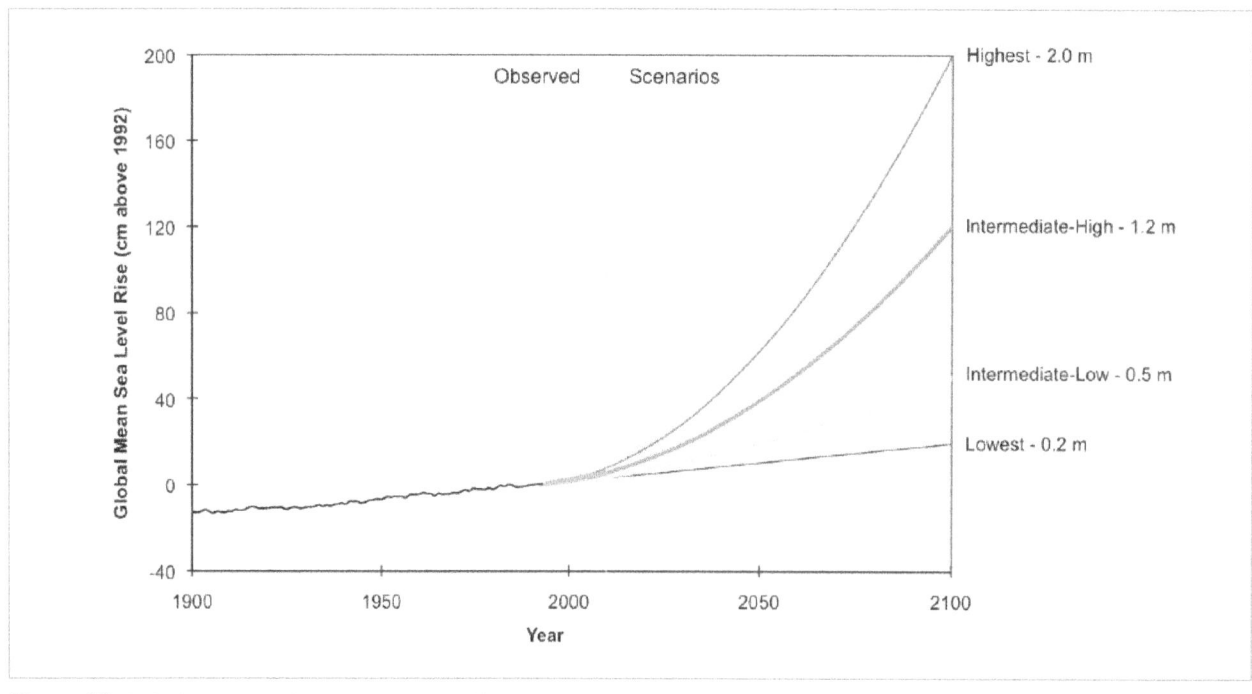

Figure 10. Global mean sea level rise scenarios. Present Mean Sea Level (MSL) for the US coasts is determined from the National Tidal Datum Epoch (NTDE) provided by NOAA. The NTDE is calculated using tide gauge observations from 1983 – 2001. Therefore, we use 1992, the mid-point of the NTDE, as a starting point for the projected curves. The Intermediate High Scenario is an average of the high end of ranges of global mean SLR reported by several studies using semi-empirical approaches. The Intermediate Low Scenario is the global mean SLR projection from the IPCC AR4 at 95% confidence interval.

The Lowest Scenario is based on a linear extrapolation of the historical SLR rate derived from tide gauge records beginning in 1900 (1.7 mm/yr). Global sea level has risen 0.2 meters (8 inches) over this period of record, and we anticipate at least another 8 inches by 2100. The rate of global mean SLR derived from satellite altimetry (1992 to 2010) has been substantially higher (3.2 mm/yr), approaching twice the rate of the longer historical record from tide gauges. However, the 18-year altimeter record is insufficient in duration for projecting century-scale global SLR. Trends derived from the shorter records are less reliable as projections because they are affected by inter-annual and decadal climate and oceanographic patterns that are superimposed upon the long-term rise of global sea level. The Lowest Scenario should be considered where there is a great tolerance for risk.

There is a highly significant correlation between observations of global mean SLR and increasing global mean temperature (Vermeer and Rahmstorf 2009, Rahmstorf et al. 2011), and the IPCC (2007a) and more recent studies (Schaeffer et al. 2012) anticipate that global mean sea level will continue to rise even if warming ceases. Our Highest Scenario is an upper limit for SLR by 2100, but the possibility exists that SLR could exceed this limit beyond this timeframe (Pfeffer et al 2008). Our Intermediate-Low and Lowest Scenarios are optimistic scenarios of future environmental change assuming rates of ice sheet loss and ocean warming slightly higher or similar to recent observations.

4.2 Ice Sheet Loss

Other studies (e.g. Rohling et al. 2008) have arrived at even greater estimates of future global mean SLR than our Highest Scenario, but we are not confident in the plausibility of those estimates at this time. The IPCC AR4 produced some of the more widely used projections of global SLR for the 21st century (IPCC 2007a). The IPCC projections included ocean warming, contributions from glaciers, and modeled partial ice sheet contributions. The IPCC AR4 estimates did not include, however, potential rapid dynamic response of Greenland and Antarctic Ice Sheets as reflected in our Highest Scenario.

A growing body of recently published work suggests that, due to increasing loss, the great polar ice sheets in Greenland and Antarctica will become much more significant contributors to global SLR in the future (e.g., Rignot et al. 2011, Vermeer and Rahmstorf 2009, Van den Broeke et al 2011, NRC 2012). Ice sheet contributions to global mean SLR stem from mass loss brought about by melting and discharge of ice into the ocean at marine-terminating glaciers and ice streams (NRC 2012). Multiple reports indicate that mass loss of both the Greenland and Antarctic ice sheets may have accelerated over the past two decades, despite high inter-annual variability in space and time (Chen et al. 2011, Rignot et al. 2011, Van den Broeke 2011, NRC 2012). For example, regional variability of mass loss from the Greenland ice sheet (GIS) over the past few years shows that areas of accelerating deterioration changed from the southeast part of the ice sheet to the northwest part, suggesting high sensitivity of the GIS to regional climate (Chen et al. 2011). In Antarctica, ice loss is occurring in some coastal areas, and ice accumulation is occurring in interior Antarctica. While the balance between ice loss and accumulation remains an area of investigation, recent observations suggest that ice loss has been greater (King et al. 2012).

Most of the ice loss in Antarctica has come from the West Antarctic ice sheet (WAIS; Rignot et al. 2008). A significant portion of the WAIS is floating at or grounded below sea level, as are relatively smaller parts of the ice sheets in East Antarctica and Greenland. Floating ice shelves support land-based ice sheets. Current and future ocean warming below the surface make ice shelves susceptible to catastrophic collapse, which in turn can trigger increased ice discharge to the ocean (Rignot et al. 2004, Scambos et al. 2004, Jacobs et al. 2011, Joughlin and Alley 2011, Yin et al. 2011). Better understanding of how the polar ice sheets will respond to further changes in climatic conditions over the 21st century requires continued development of physical models (Price et al. 2011).

Ice sheet losses will lower the gravitational attraction ice sheets have for surrounding seas, producing spatial variability in changes to global mean sea level

Main Report

(Kopp et al. 2010; Mitrovica et al. 2001; Mitrovica et al. 2009). Although seemingly counterintuitive, sea level falls close to deteriorating ice sheets even though ice sheet losses are discharged into the adjacent sea. This lowering of sea level is due to gravitational effects that can cause SLR up to ~2,000 km from the melting ice sheet. SLR resulting from deterioration of the GIS is thought to be relatively lower than the global average for the contiguous US, Alaska, and US territories in the Caribbean Sea and relatively higher for Hawaii and US territories in the Pacific Ocean (Kopp et al. 2010). SLR resulting from deterioration of the WAIS is thought to be relatively higher than the global average for all states and territories of the US (Mitrovica et al. 2009).

4.3 Methods for Constructing Scenario Data

Both the rate and magnitude of SLR are important for vulnerability and impact assessment because, as mentioned below, the time horizon is a critical factor affecting risk tolerance for coastal management actions. Even long-term coastal management actions (e.g. coastal habitat restoration) are sensitive to near-term rates and amounts of SLR. To address these considerations, we provide the following methodology for creating curves, anchored to a specific date, and yielding estimates for specific time horizons.

Future estimates for global mean SLR are relative to the current elevation of global mean sea level. It is important to select a starting point in time from which to move forward with the scenarios. Present Mean Sea Level (MSL) for the US coasts is determined from long-term NOAA tide gauge records and is referenced to the current National Tidal Datum Epoch (NTDE) provided by NOAA. The NTDE is a 19-year period with the current NTDE being 1983 to 2001. NOAA uses the NTDE as the basis for all tidal datums (i.e., Mean High Water and Mean Lower Low Water) and uses NTDE MSL as the reference for presentation of RSL trends (Figure 3). MSL for the NTDE is the mean of hourly heights observed over the entire 19-year period. Because mean sea level is an average over the 19-year NTDE, the mean sea level value is

associated with the mid-point of the NTDE, which is the year 1992. As the mid-point for the NTDE, 1992 is selected as the start-point for our scenarios (Figure 10, also see Flick et al. 2012).

As described above, the Lowest Scenario is a linear extrapolation of the historic trend of 20th century tide gauge measurements. In contrast, the Highest, Intermediate-High and Intermediate-Low Scenarios represent possible future acceleration in global mean sea level rise and thus are not described by linear relationships. To represent the non-linear trajectory of SLR in these scenarios, the NRC (1987) and USACE (2011) scheme is adopted, wherein the future global mean SLR is represented by the following quadratic equation:

$$E(t) = 0.0017t + bt^2 \qquad (1)$$

in which t represents years, starting in 1992, b is a constant, and $E(t)$ is the eustatic SLR, in meters, as a function of t. To fit the curves to our scenarios, as defined above, the constant b has a value of 1.56E-04 (Highest Scenario), 8.71E-05 (Intermediate-High Scenario), and 2.71E-05 (Intermediate-Low Scenario). It should be emphasized that this straightforward quadratic approach to the time evolution is chosen in part for its simplicity; there is no scientific reason or evidence to assume that SLR will evolve in precisely this smooth manner.

If it is necessary to estimate a projected rise in global mean sea level for any of the scenarios, but starting in a year more recent than 1992, the following equation can be used:

$$E(t2) - E(t1) = 0.0017(t2 - t1) + b(t2^2 - t1^2) \qquad (2)$$

where t1 is the time between the beginning year of interest and 1992 and t2 is the time between the ending year of interest and 1992 (Knuuti 2002, Flick et al. 2012). For example, global mean SLR for any of the accelerating scenarios can be calculated over the period of time between 2011 and 2050 by using t1 = 2011-1992 = 19 years and t2 = 2050-1992 = 58 years.

5. Enhancing the Global SLR Scenarios

The development of sea level change scenarios (global, regional, and local) is an initial stage in conducting coastal vulnerability assessments (Figure 11). We recommend that the choice of scenarios involve interdisciplinary scientific experts, as well as coastal managers and planners who understand relevant decision factors. Three decision factors generally considered in the choice of sea level scenarios are: location, time horizon, and risk tolerance (Mote 2008).

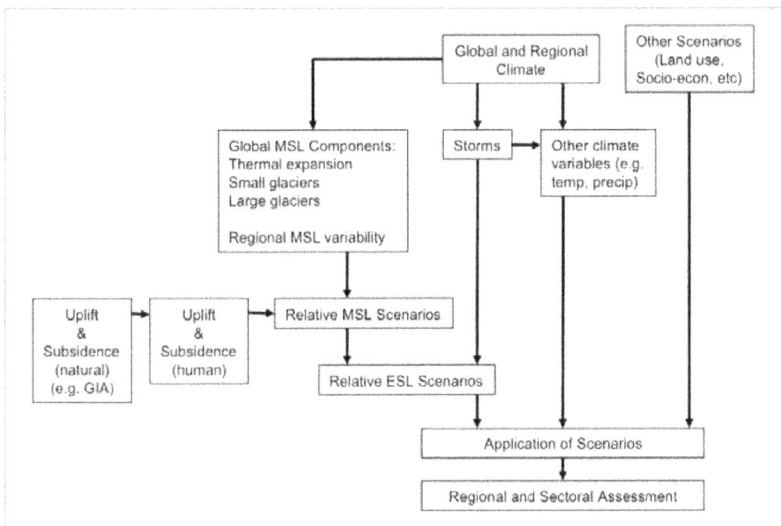

Figure 11. Developing sea level scenarios (Nicholls et al 2011). MSL – mean sea level; ESL – Extreme sea level.

Location refers to the planning or management area for which assessment or analysis is being conducted. Time horizon refers to the committed life span of a particular use in a coastal area and varies depending on whether you are planning temporary flood control, building long-term infrastructure such as an airport or shipyard, or restoring or preserving ecosystem function. Finally, risk tolerance refers to a community's or decision maker's willingness to accept a higher or lower probability of impacts.

Risks associated with SLR may not be evident when considering sea level change in isolation from climate or over a narrowly defined coastal planning area. Power stations or airports at specific locations along the coast may be critically important to the regional or national

economy and thus may be protected with a low tolerance for projected long-term, regional, or global scale impacts (e.g. a large levee). Such levels of protection, however, may have adverse effects on adjacent parts of the coast or create a false sense of reduced risk if sea level rises and coastal flooding increases (Smits et al. 2006, Parris and Lacko 2009).

Additional information should be combined with the global scenarios to incorporate regional and local conditions when conducting risk analysis. These factors include regional mean sea level variability, local and regional vertical land movement, coastal environmental processes (geological, ecological, biological, and socio-economic), and the effect of extreme weather and climate on RSL. Much of this information is being prepared for the NCA, including scenarios of climate, land-use/land change, and different socio-economic conditions as well as analysis of changes in extreme weather and climate. The NOAA Coastal Services Center and the USGS also provide access to information via Digital Coast, including two companion reports on developing sea level scenarios (NOAA 2012, NOAA 2010). As mentioned previously, regional and local sea level scenarios are beyond the scope of this report. We provide Table 3 as a basic template for those people who wish to build on our scenarios.

5.1 Vertical Land Movement (VLM)

Regional and local sea level scenarios should account for vertical land movement. Following the last glacial maximum, the Earth's crust[1] began to rise in response to earlier subsidence under the weight of the ice. This process, post-glacial rebound, varies across the continent and may be a significant factor in vertical land movement in some areas. It is typically accounted for by estimating GIA.

[1] Continental crust – The crust is the solid, outermost layer of the Earth that makes up the continents (continental crust), including mountain ranges, and that lies beneath the oceans (oceanic crust).

Main Report

GIA refers to the rise of the Earth's crust that was depressed by the weight of continental ice sheets during the last glacial maximum (Appendix 1; Lambeck et al. 2010, Mitrovica et al. 2010). The reaction of the Earth's crust to ice load changes is a relatively slow and delayed process dependent upon the viscosity of the mantle, on which the Earth's crust rests. This reaction continues long after ice sheets melt and water volumes stabilize. For example, the continental crust is still at rates of mm-per-year. As a result, it is necessary to consider GIA in the analysis of sea level observations and trends. Recent progress in GIA models has reduced uncertainties in ice load history and mantle viscosity, improving interpretations of sea level analyses (Mitrovica et al. 2010, Peltier 2004).

The contribution of GIA to net changes in RSL varies for much of the US coastline and its island territories, with modeled rates ranging from +0.5 to +2.0 mm/yr for much of the mainland and northern Alaska (Figure 12). Portions of the Northeast and Northwest mainland coasts have modeled rates at the upper end of this range. Rates vary from slightly positive to slightly negative for only the most southern and northern latitudes of the mainland, as well as for much of Alaska and the island states and territories. Southern Alaska is the only region that shows locations with a falling RSL due to GIA effects, with modeled rates between -0.5 and -1.7 mm/yr at a few sites.

Table 3. Template for developing regional or local sea level scenarios

Contributing Variables	Scenarios of Sea Level Change			
	Lowest Scenario	Intermediate-Low Scenario	Intermediate-High Scenario	Highest Scenario
Global mean sea level rise* (m)	0.2	0.5	1.2	2.0
Vertical Land Movement (subsidence or uplift)**				
Ocean Basin Trend*** (from tide gauges and satellites)				
Total Relative Sea Level Change				
Extreme Water Level (from existing flood models or long-term tide gauges)				

* Equations from Section 4.3 can be used to calculate scenarios of sea level change over desired period and to populate the global mean SLR term in the first row.

** This row can be populated using, in part, the information found in Sections 5.1.

*** This row can be populated using, in part, the information found in Sections 3.1, 3.2, 3.3 and 5.3.

experiencing post-glacial rebound from the reduction of ice sheets present during the last glacial maximum, which occurred approximately 21,000 years ago.

Surface deformations from VLM induced by GIA can raise or lower RSL, depending on the proximity to the prior ice sheet (Lambeck et al. 2010, Mitrovica et al. 2010). Glacial rebound in North America is largest around Hudson Bay, where the former ice sheet was both centered and thickest. Meanwhile, land is subsiding further south at and beyond the edges of the former ice sheet, in response to the rebound to the north. GIA is changing coastal land elevations

With the advent of Global Positioning Systems (GPS), precise spatial estimates of vertical and horizontal land movement are becoming available. Using over three hundred and sixty GPS sites in Canada and United States, Sella et al. (2007) produced a map of "vertical velocities" which may be used to understand the rates of VLM that vary within an order of magnitude along much of the coastal region of the east and the south (Figure 13). Most of the GPS measurements demonstrated consistency with GIA models. The magnitude of subsidence along the eastern US is about 1 to 2 mm. The international community has recognized

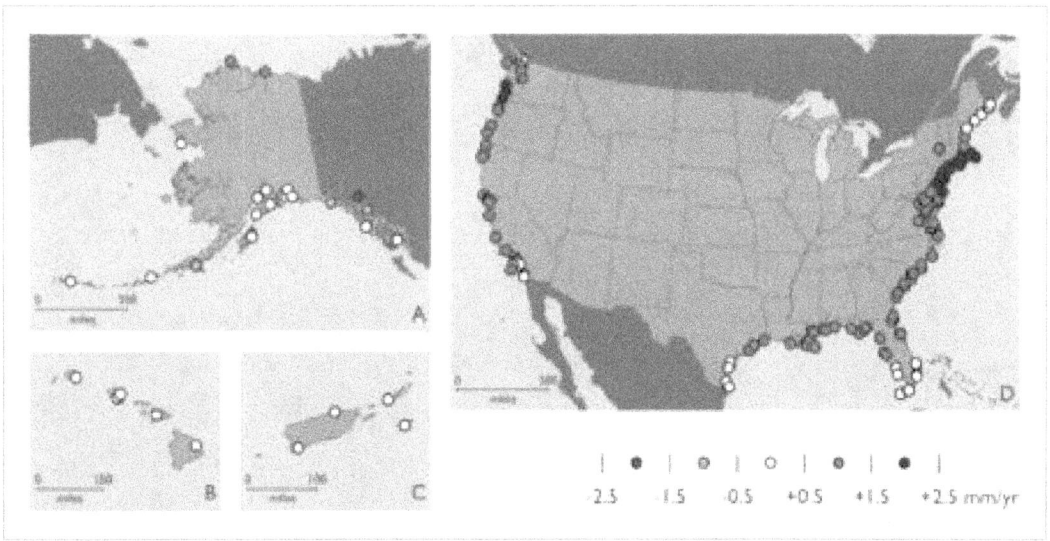

Figure 12. Contribution of GIA to net changes in RSL varies for much of the US coastline Alaska (A), Hawaii (B), Puerto Rico and the US Virgin Islands (C), and the contiguous US (D), as predicted by the ICE-5G(VM2) model (Peltier 2004). Rate locations are tide gauge sites in the Permanent Service for Mean Sea Level (PSMSL) dataset. Values account for sharp coastal gradients through use of a high spatial resolution model, and represent rates averaged over the 20th and 21st centuries. For US territories not shown, rates fall within the -0.5 to +0.5 mm/yr range.

the need to upgrade the observational infrastructure to include co-location of continuous GPS receivers and tide gauges (see Continuous GPS Update). Nationally, NOAA has recognized the need to co-locate these systems as much as possible and has already integrated repeat static GPS measurements on tidal benchmarks as part of the operation and maintenance of the National Water Level Observation Network (NWLON).

At some locations along the coastline, VLM may be exacerbated by factors other than GIA. Other factors of VLM include, but are not limited to: plate tectonics; natural compaction of thick layers of loose, unconsolidated sediments; sediment compaction due to excessive withdrawal of groundwater, oil, or gas; and subsidence due to oxidation of organic soil. Sun et al. (1998) indicate that land subsidence due to groundwater withdrawal in southern New Jersey has increased from 2 cm to 3 cm over the past 40 years. Along the Gulf coast states, subsidence is widespread extending beyond the Mississippi Delta to Mississippi and the estimated rates are as high as 25 mm/year with much of the delta subsiding at 5 to 10 mm/yr (Gonzalez and Tornqvist 2006, Shinkle and Dokka 2004). In Alaska, the uplift rates are much higher as

they are caused by significant tectonic activity in the region as well as post-glacial rebound (Larsen et al. 2003, Larsen et al. 2007).

Recent geodetic studies clearly document non-GIA VLM and/or RSL trends over spatial scales approaching 100s of kilometers. For example, traditional geodesy and continuous global positioning systems (GPS)

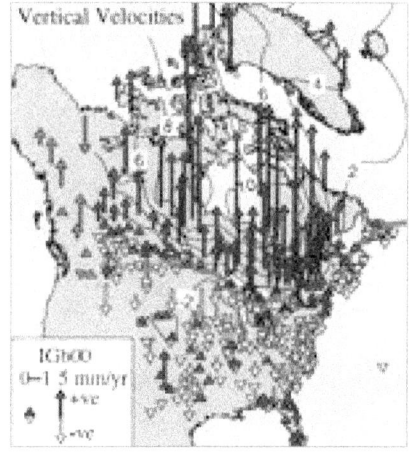

Figure 13. Vertical GPS site motions with respect to IGb00 (a reference plane) as estimated by Sella et al. (2007).

Main Report

studies in the Cascadia subduction zone (Cascadia) that spans CA, OR, and WA suggest regional uplift signals in the range of 1 to 3mm/yr (Burgette et al. 2009, Mazzotti et al. 2007). Whether this rate is constant over time depends on earthquake variability in Cascadia (Wang et al. 2003, Wang 2007). However, modern, observable rates in Cascadia essentially negate global mean SLR in Alaska and parts of Washington and Oregon. More detailed VLM data are provided for specific areas in California, including the San Francisco Bay area and the Los Angeles Basin (Burgmann et al. 2005; Brooks et al. 2007). Deep subsidence in the Gulf of Mexico has been linked to a combination of groundwater withdrawal, regional tectonic loading on the Earth's crust from the Mississippi River Delta, and possibly faulting (Dokka 2011).

5.2 Extremes of Weather and Climate

Coastal managers are most immediately concerned with the effect that global, regional, and RSL changes have on coastal flooding. In addition to changes in mean sea level, consideration of future extremes is vitally important for planning and design of coastal infrastructure (Zhang et al. 2000). SLR can amplify factors that currently contribute to coastal flooding: high tides, storm surge, high waves, and high runoff from rivers and creeks (Cayan et al. 2008). All of these factors change during extreme weather and climate events and often have disproportionately high impacts in most coastal regions. Although a consensus has not yet been reached on how the frequency and magnitude of storms may change in coastal regions of the US, it is certain that higher mean sea levels increase the frequency, magnitude, and duration of coastal flooding associated with a given storm. Thus, considering the impact of different weather events combined with scenarios of sea level change is crucial in developing hazard profiles for emergency planning and vulnerability, impact, and adaptation assessments.

Several studies indicate that the number and duration of extreme high-water events increase during Atlantic Multidecadal Oscillation (AMO) warm phases around the Florida Peninsula (Park et al 2010a, Park et al 2010b, Park et al 2011). As mentioned previously, sea level anomalies related to ENSO along the eastern Pacific coast can result in as much as 20 cm of RSL rise for an entire winter season (a period of high tides, storms, and large waves) (Bromirski and Flick 2008). In the San Francisco Bay, global mean SLR and the ENSO related anomaly are projected to increase the number of sea level extremes (Cayan et al. 2008).

Coastal vulnerability assessments in some urban centers of the US have incorporated SLR and climate scenarios into probabilistic flood projections. For example, the return period concept used in flood control projects provides a risk-based approach to select design flood conditions based on probability of occurrence (Horton et al 2010, Knowles 2010). The occurrence of storm surges and high tides have been incorporated into a Joint Probability Method (JPM) in an effort to provide probabilistic projections of extreme levels, but the JPM methods do not account for future changes in sea level and runoff due to climate change (Tomasin and Pirazzoli 2008, Liu et al. 2010). These methodologies may be particularly useful for incorporating sea level into a risk management framework as part of any assessment process.

5.3 Dynamic Changes in Ocean Circulation

Regional and local sea level change scenarios also should reflect dynamic changes in ocean circulation, ice sheet losses, and mass redistributions associated with ice sheet losses, though this information is very limited at present. Climate models have forecast the slowing of the overturning circulation in the Atlantic from warming seawater in the north Atlantic, introduction of fresh water from GIS, and other sources. These factors could slow the boundary currents along the US east coast and raise sea levels in the northeast (Yin et al. 2009, Yin et al. 2010). These changes are not entirely independent from ice sheet loss, and can exacerbate or attenuate rises in sea level, depending on the region. Climate models forced by the SRES A1B scenario reveal a potential slowing in the Atlantic meridional overturning circulation (AMOC) and regional thermosteric and halosteric changes (Yin et al. 2010). Dynamical SLR resulting from ocean circulation patterns could be additive to the global mean SLR trend, creating even higher sea levels and potential coastal impacts in Boston, New York, and Washington, DC when compared to the Southeastern US (Yin et al. 2010).

5.4 Other stressors related to Coastal Vulnerability

Coastal land use and landforms affect inundation patterns on the coast. In addition to the aforementioned factors, a number of local coastal processes determine the configuration of coastal landforms (e.g. beach profiles, sea cliffs, barrier islands, marshes, atolls, etc.; see CCSP 2009). Specific locations where these factors dominate coastal processes are difficult to integrate into scenarios of environmental change (Cloern et al 2011). They should at least be considered in determining vulnerability and impacts. These factors vary substantially among US coastal areas and include, but are not limited to:

- Sediment supply to the coast and associated transport along the coast
- Elevation and range of tides
- Wave height, period, and slope of the shoreline
- Sediment accumulation rates – physical and biogeochemical
- Presence or absence and configuration of barriers, whether human-made or natural, to coastal flooding and to constraints on shoreline change
- Permafrost decline – a unique driver of coastal elevation change in northern Alaska that controls, in some locales, the retreat of coastal bluffs.

6. Conclusions

Based on a large body of science, we identify four scenarios of global mean SLR ranging from 0.2 meters (8 inches) to 2.0 meters (6.6 feet) by 2100. These scenarios provide a set of plausible trajectories of global mean SLR for use in assessing vulnerability, impacts, and adaptation strategies. None of these scenarios should be used in isolation, and experts and coastal managers should factor in locally and regionally specific information on climatic, physical, ecological, and biological processes and on the culture and economy of coastal communities. Scientific observations at the local and regional scale are essential to action, but global phenomena, such as SLR, can influence those conditions creating unanticipated impacts at the local scale, especially over longer time horizons. Thousands of structures along the US coast are over fifty years old including vital storm and waste water systems. Thus, coastal vulnerability, impact, and adaptation assessments require an understanding of the long-term, global and regional drivers of environmental change.

References

Ablain, M., A. Cazenave, et al. (2009). "A new assessment of the error budget of global mean sea level rate estimated by satellite altimetry over 1993-2008." Ocean Science 5(2): 193-201.

Becker, M., B. Meyssignac, C. Letetrel, W. Llovel, A. Cazenave, and T. Delcroix (2012) Sea level variations at tropical Pacific islands since 1950, Global and Planetary Change, 80-81, 85-98.

Berger WH (2008) Sea level in the late Quaternary: patterns of variation and implications. Int J Earth Sci 97:1143-1150, doi:10.1007/s00531-008-0343-y

Boon, John D. (2012) "Evidence of Sea Level Acceleration at U.S. and Canadian Tide Stations, Atlantic Coast, North America" Journal of Coastal Research, 28(6) 1437:1445.

Bromirski, P., R.E. Flick and D.R. Cayan, 2003: Storminess Variability Along the California Coast: 1858-2000. J. Climate, 16(6), 982-993.

Bromirski, P., and R.E. Flick 2008. Storm surge in the San Francisco Bay/Delta and nearby coastal locations. Shore and Beach, vol. 76, no. 3 pp 29 - 37.

Bromirski, P.D., A.J. Miller, R.E. Flick, and G. Auad (2011). Dynamical suppression of SLR along the Pacific coast of North America: Indications for imminent acceleration, J. Geophys. Res. – Oceans, 116, C07005, doi:10.1029/2010JC006759.

Brooks, B., M. A. Merrifield, J. Foster, C. L. Werner, F. Gomez, M. Bevis, and S. Gill (2007), Space geodetic determination of spatial variability in relative sea level change, Los Angeles Basin, Geophys. Res. Lett., 34, L01611, doi:10.1029/2006GL028171.

Burgette, R.J., Weldon II, R.J., and D.A. Schmidt. 2009. Interseismic uplift rates for western Oregon and along-strike variation in locking on the Cascadia subduction zone. JOURNAL OF GEOPHYSICAL RESEARCH, VOL. 114, B014

Bürgmann, R., Hilley, G., Ferretti, A., and F. Novali. 2005. Resolving vertical tectonics in the San Francisco Bay Area from permanent scatterer InSAR and GPS analysis. Geology v. 34 no. 3 p. 221-224

Cayan, D.R., P.D. Bromirski, K. Hayhoe, M. Tyree, M..D. Dettinger and R.E. Flick, 2008: Climate change projections of sea level extremes along the California coast. Climatic Change, 87, (Suppl 1):S57-S73, doi:10.1007/s10584-007-9376-7.

CCSP, 2009: Coastal Sensitivity to Sea-Level Rise: A Focus on the Mid-Atlantic Region. A report by the US Climate Change Science Program and the Subcommittee on Global Change Research. [James G. Titus (Coordinating Lead Author), K. Eric Anderson, Donald R. Cahoon, Dean B. Gesch, Stephen K. Gill, Benjamin T. Gutierrez, E. Robert Thieler, and S. Jeffress Williams (Lead Authors)]. US Environmental Protection Agency, Washington D.C., USA, 320 pp.

Chen JL, Wilson CR, Tapley BD (2011) Interannual variability of Greenland ice losses from satellite gravimetry. J Geophys Res 116:B07406 doi:10.1029/2010JB007789

Church, J. A. and N.J. White. 2011. Sea-level rise from the late 19th to the early 21st Century. Surveys in Geophysics, doi:10.1007/s10712-011-9119-1

Cloern JE, Knowles N, Brown LR, Cayan D, Dettinger MD, et al. (2011) Projected Evolution of California's San Francisco Bay-Delta-River System in a Century of Climate Change. PLoS ONE 6(9): e24465. doi:10.1371/journal.pone.0024465

Crowell, M, Coulton, K, Johnson, C, Westcott, J, Bellomo, D Edelman, S and E Hirsch. 2010. An Estimate of the U.S. Population Living in 100-Year Coastal Flood Hazard Areas. Journal of Coastal Research, v. 26, no. 2, 201 – 211.

References

Dokka, R. 2011. The role of deep processes in late 20th century subsidence of New Orleans and coastal areas of southern Louisiana and Mississippi. JOURNAL OF GEOPHYSICAL RESEARCH, VOL. 116, B06403, 25 PP

Feng, M., M. J. McPhaden, and T. Lee (2010), Decadal variability of the Pacific subtropical cells and their influence on the southeast Indian Ocean, Geophys. Res. Lett., 37, L09606, doi:10.1029/2010GL042796

Flick E. R, Knuuti, K., and Gill, S.K. (2012) Matching Mean Sea Level Rise Projections to Local Elevation Datums. J. Waterway, Port, Coastal, Ocean Eng. doi: 10.1061/(ASCE)WW.1943-5460.0000145

González, J.L. and Törnqvist, T.E.. 2006. Coastal Louisiana in crisis: Subsidence or sea level rise? Eos, 87: 493, 498.

Gray, S. 2011. From Uncertainty to Action: Climate Change Projections and the Management of Large Natural Areas. BioScience 61(7): 504-505.

Grinsted, A., J. C. Moore, and S. Jevrejeva (2009), Reconstructing sea level from paleo and projected temperatures 200 to 2100AD, Clim. Dyn., doi:10.1007/s00382-008-0507-2.

Hayden, T., Kominz, M., Powars, D., Edwards, L., Miller, K., Browning, J., and Kulpecz, A. (2008) Impact effects and regional tectonic insights: Backstripping the Chesapeake Bay impact structure. Geology. v. 36, p. 327-330

Hearty, P. J., J. T. Hollin, A. Neumann, M. O'Leary, and M. McCulloch (2007), Global sea-level fluctuations during the last interglaciation (MIS 5e), Quat. Sci. Rev., 26, 2090–2112, doi:10.1016/j.quascirev. 2007.06.019.

Henrichs, T., M. Zurek, B. Eickhout, K. Kok, C. Raudsepp-Hearne, T. Ribeiro, D. van Vuuren, nd A. Volkery. 2010. "Scenario development and analysis for forward-looking ecosystem assessments." In Ecosystems and human well-being: A manual for assessment practitioners, edited by Neville Ash, Hernan Blanco, Keisha Garcia, Thomas Tomich, Bhaskar Vira, Monika B. Zurek and Claire Brown.

Horton, R. M., V. Gornitz, D. A. Bader, A. C. Ruane, R. Goldberg, and C. Rosenzweig, 2011: Climate Hazard Assessment for Stakeholder Adaptation Planning in New York City. J Appl Meteorol Clim, 50, 2247-2266, doi:10.1175/2011JAMC2521.1.

Horton, R, Gornitz, V, Bowman, M, and Blake, R (2010) Chapter 3: Climate observations and projections in Annals of the New York Academy of Sciences - Volume 1196 Climate Change Adaptation in New York City: Building a Risk Management Response: New York City Panel on Climate Change 2010 Report

Horton, Radley, Herweijer, Celine, Rosenzweig, Cynthia, Liu, Jiping, Gornitz, Vivien and Ruane, Alex C. (2008), Sea level rise projections for current generation CGCMs based on the semi-empirical method, Geophysical Research Letters, Vol. 35, L02715, doi: 10.1029/2007GL032486, 2008

Houston, J. R., and R. G. Dean (2011). Sea-level acceleration based on US tide gages and extensions of previous global-gage analyses, J. Coastal Res., 27, 409-417.

http://ro.uow.edu.au/cgi/viewcontent.cgi?article=1251&context=scipapers

Intergovernmental Panel on Climate Change (IPCC) (2000) Special Report on Emissions Scenarios. (Nakicenovic, N., and R. Swart. eds.). Cambridge University Press, Cambridge, United Kingdom. http://www.grida.no/climate/ipcc/emission/

References

IPCC (2001) Climate Change 2001: the scientific basis. Contribution of Working Group 1 to the Third Assessment Report of the Intergovernmental Panel on Climate Change, edited by J. T. Houghton, Y. Ding, D. J. Griggs, M. Noguer, P. J. van der Linden, X. Dai, K. Maskell and C. A. Johnson (eds). Cambridge University Press, Cambridge, UK, and New York, USA, 2001.

IPCC (2007a) Climate Change 2007: The Physical Science Basis. Contribution of Working Group I to the Fourth Assessment Report of the Intergovernmental Panel on Climate Change [Solomon, S., D. Qin, M. Manning, Z. Chen, M. Marquis, K.B. Averyt, M.Tignor and H.L. Miller (eds.)]. Cambridge University Press, Cambridge, United Kingdom and New York, NY, USA.

IPCC (2007b) IPCC Fourth Assessment Report Annex 1: Glossary. In: Climate Change 2007: The Physical Science Basis. Contribution of Working Group I to the Fourth Assessment Report of the Intergovernmental Panel on Climate Change (Solomon, S., D. Qin, M. Manning, Z. Chen, M. Marquis, K. B. Averyt, M. Tignor, and H. L. Miller, eds.). Cambridge University Press, Cambridge, United Kingdom and New York, NY, USA.

Jacobs SS, Jenkins A, Giulivi CF, Dutrieux P (2011) Stronger ocean circulation and increased melting under Pine Island Glacier ice shelf. Nat Geosci 4:519-523

Jevrejeva S, Moore JC, Grinsted A. 2010. How will sea level respond to changes in natural and anthropogenic forcings by 2100? Geophysical Research Letters 37

Jevejeva, S., J.C. Moore, A. Grinsted (2012) Se a Projections to AD2500 with a new generation of climate change scenarios, Global and Planetary Change, 80-81, 14-20.

Joughin, I., and R.B. Alley. 2011. Stability of the West Antarctic ice sheet in a warming world. Nature Geoscience 4, 506–513.

Katsman, CA, Sterl, A, Beersma, JJ, van den Brink, HW, Church, JA, Hazeleger, W, Kopp, RE, Kroon, D, Kwadijk, J, Lammersen, R, Lowe, J, Oppenheimer, M, Plag, H-P, Ridley, J, von Storch, H, Vaughan, DG, Vellinga, P, Vermeersen, LLA, van de Wal, RSW & Weisse, R 2011, 'Exploring high-end scenarios for local sea level rise to develop flood protection strategies for a low-lying delta-the Netherlands as an example' Climatic Change, vol 109, no. 3-4, pp. 617-645.

Kawamura, K., Parrenin, F., Lisiecki, L., Uemura, R., Vimeux, F., Severinghaus, J.P., Hutterli, M., Nakazawa, T., Aoki, S., Jouzel, J., Raymo, M., Matsumoto, K., Nakata, H., Motoyama, H., Fujita, S., Goto-Azuma, K., Fujii, Y., and Watanabe, O. 2007. Northern Hemisphere forcing of climatic cycles in Antarctica over the past 360,000 years. Nature 448: 912-916

Kemp, A. C., Horton, B.P., Donnelly, J.P., Mann, M.E., Vermeer, M., Rahmstorf, S. 2011. Climate related sea-level variations over the past two millennia. Proceedings of the National Academy of Sciences. DOI 10.1073 PNAS 1015619108

King, M.A., R.J. Bingham, O. Moore, P.L. Whiehouse, M.J. Bentley, and G.A. Milne (2012). Lower satellite-gravimetry estimates of Antarctic sea-level contribution, Nature, doi:10.1038/nature11621.

Knowles, Noah. 2010. Potential Inundation Due to Rising Sea Levels in the San Francisco Bay Region. California Climate Change Center. San Francisco Estuary and Watershed Science, 8:1.

Knuuti, K. (2002) Planning for Sea-Level Rise: U.S. Army Corps of Engineers Policy. Solutions to Coastal Disasters '02, ASCE 2002

Kopp. R., F. J. Simons, J. X. Mitrovica, A. Maloof, and M. Oppenheimer (2009), Probabilistic assessment of sea level during the Last Interglacial stage, Nature, 462, 863–867, doi:10.1038/nature08686.

Kopp RE, Mitrovica JX, Griffies, Yin J, Hay CC, Stouffer RJ (2010) The impact of Greenland melt on local sea levels: a partially coupled analysis of dynamic and static equilibrium effects in idealized water-hosing experiments. Clim Change 103:619-625

Lambeck K., Woodroffe C.D., Antonioli F, Anzidei M, Gehrels WR, Laborel J, Wright AJ (2010) Paleoenvironmental records, geophysical modeling, and reconstruction of sea-level trends and variability on centennial and longer timescales. In: Church JA, Woodworth PL, Aarup T, Wilson WS (eds) Understanding sea-level rise and variability. Blackwell Publishing Ltd, pp 61-121

Larsen, C. F., K. A. Echelmeyer, J. T. Freymueller, and R. J. Motyka (2003) Tide gage records of uplift along the northern Pacific-North American plate boundary, 1937 to 2001, 2003. J. Geophys. Res., 108(B4), 2216, doi:10.1029/2001JB001685.

Larsen, C.F., R.J. Motyka, A.A. Arendt, K.A. Echelmeyer, and P.E. Geissler (2007) Glacier changes in southeast Alaska and northwest British Columbia and contribution to sea level rise. Journal of Geophysical Research 112. doi:10.1029/2006JF000586

Liu J. C., Lence B. J., Isaacson M. (2010), Direct Joint Probability Method for Estimating Extreme Sea Levels, Journal of Waterway, Port, Coastal, and Ocean Engineering, 136 (1), 66-76, DOI: 10.1061/ASCE0733-950X(2010)136:1(66).

Maine Department of Environmental Protection (MDEP) (2009) People and Nature Adapting to a Changing Climate: Charting Maine's Course. Accessed at: http://www.maine.gov/dep/oc/adapt/Report_final.pdf. Accessed October 27, 2011

Mantua. Nathan J. et al.1997. A Pacific interdecadal climate oscillation with impacts on salmon production. Bulletin of the American Meteorological Society 78 (6): 1069–1079.

Maryland Commission on Climate Change (MCCC) (2008) Maryland Climate Action Plan. Accessed at: www.mdclimatechange.us/ewebeditpro/items/O40F14798.pdf. Accessed on October 27, 2011.

Mazzotti, S., A. Lambert, N. Courtier, L. Nykolaishen, and H. Dragert (2007), Crustal uplift and sea level rise in northern Cascadia from GPS, absolute gravity, and tide gauge data, Geophys. Res. Lett., 34, L15306, doi:10.1029/2007GL030283.

McKay NP, JT Overpeck, BL Otto-Bliesner (2011) The role of ocean thermal expansion in Last Interglacial sea level rise. Geophysical Research Letters, 38, L14605, doi:10.1029/2011GL048280

McKenzie, E., A. Rosenthal et al. 2012. Developing scenarios to assess ecosystem service tradeoffs: Guidance and case studies for InVEST users. World Wildlife Fund, Washington, D.C.

Meehl, G.A., T.F. Stocker, W.D. Collins, P. Friedlingstein, A.T. Gaye, J.M. Gregory, A. Kitoh, R. Knutti, J.M. Murphy, A. Noda, S.C.B. Raper, I.G. Watterson, A.J. Weaver and Z.-C. Zhao, 2007. Global Climate Projections. In Solomon et al. (Eds.), Climate Change 2007: The Physical Science Basis. Contribution of Working Group I to the Fourth Assessment Report of the Intergovernmental Panel on Climate Change. Cambridge University Press, Cambridge. United Kingdom and New York. NY, US, pp. 749-845.

Merrifield, M.A. (2011). A shift in western tropical Pacific sea level trends during the 1990s, J. Clim., 24(15), 4126-4138.

References

Merrifield, M. A., and M. E. Maltrud, (2011) Regional sea level trends due to a Pacific trade wind intensification, Geophys. Res. Lett., 38, L21605, doi:10.1029/2011GL049576.

Mitchum, G. (2011). Sea level changes in the Southeastern United States – Past, Present and Future. A report of the Florida Climate Institute and the Southeast Climate Consortium.

Mitrovica JX, Tamisiea ME, Ivins ER, Vermeersen LLA, Milne GA, and Lambeck K (2010) Surface mass loading on a dynamic Earth: complexity and contamination in the geodetic analysis of global sea-level trends. In: Church JA, Woodworth PL, Aarup T, Wilson WS (eds) Understanding sea-level rise and variability. Blackwell Publishing Ltd, pp 285-325

Mitrovica, J.X., Gomez, N., Clark, P., 2009, The Sea-Level Fingerprint of West Antarctic Collapse, Science, Vol 323, p. 753.

Mitrovica, J. X. and G. A. Milne. 2002. On the origin of late Holocene sea-level highstands within equatorial ocean basins. Quat. Sci. Rev., 21, 2179–2190.

Mitrovica, J.X., Tamisiea, M, Davis, J. and Milne, G., 2001, Recent mass balance of polar ice sheets inferred from patterns of global sea-level change, Nature, Vol. 409.

Moss, R.H. and G. Yohe. 2011. Assessing and Communicating Confidence Levels and Uncertainties in the Main Conclusions of the NCA 2013 Report: Guidance for Authors and Contributors. National Climate Assessment Development and Advisory Committee (NCADAC).

Moss, R., Edmonds, J. A., Hibbard, K.A., Manning, M.R., Rose, S.K., van Vuuren, D. P., Carter, T., Emori, S., Kainuma, M., Kram, T., Meehl, G. A., Mitchell, G. F. B., Nakicenovic, N., Riahi, K., Smith, S.J., Stouffer, R.J., Thomson, A.M., Weyant, J.P., and T.J. Wilbanks. 2010. The next generation of scenarios for climate change research and assessment. Nature 463, 747-756 (11 February 2010) | doi:10.1038/nature08823

Mote, P., Petersen, A., Reeder, S., Shipman, H. and Binder, L. (2008) Sea Level Rise in the Coastal Waters of Washington State. A report by the University of Washington – Climate Impacts Group and the Washington Department of Ecology.

National Research Council (NRC). 1987. Responding to Changes in Sea Level: Engineering Implications. National Academy Press: Washington, D.C.

NRC. 2011. Climate Stabilization Targets: Emissions, Concentrations, and Impacts over Decades to Millennia. National Academy Press: Washington, D.C.

NRC. 2012. Sea-Level Rise for the Coasts of California, Oregon, and Washington: Past, Present, and Future. Washington, DC: The National Academies Press.

New York City Panel on Climate Change (NPCC). 2010. Climate Change Adaptation in New York City: Building a Risk Management Response. C. Rosenzweig & W. Solecki, Eds, prepared for use by the New York City Climate Change Adaptation Task Force, Annals of the New York Academy of Sciences, New York, NY. 349 pp.

Nicholls, R.J., Hanson, S.E., Lowe, J.A., Warrick, R.A., Lu, X., Long, A.J. and Carter, T.R. 2011. Constructing Sea-Level Scenarios for Impact and Adaptation Assessment of Coastal Area: A Guidance Document. Supporting Material, Intergovernmental Panel on Climate Change Task Group on Data and Scenario Support for Impact and Climate Analysis (TGICA), 47 pp.

National Oceanic and Atmospheric Administration (NOAA) National Ocean Service (NOS). 2012. Incorporating Sea Level Change Scenarios at the Local Level. NOAA NOS Technical Report. Silver Spring, MD: NOAA NOS.

National Oceanic and Atmospheric Administration (NOAA) National Ocean Service (NOS). 2010. Technical Considerations for Use of Geospatial Data in Sea Level Change Mapping and Assessment. NOAA NOS Technical Report. Silver Spring, MD: NOAA NOS.

Otto-Bliesner BL, SJ Marshall, JT Overpeck, GH Miller, A u, CAPE Last Interglacial Project members (2006) Simulating Arctic Warmth and Icefield Retreat in the Last Interglaciation. Science 311:1751-1753

Overpeck JT, Otto-Bliesner BL, Miller GH, Muhs DR, Alley RB, Kiehl JT (2006) Paleoclimatic evidence for future ice-sheet instability and rapid sea-level rise. Science 311:1747-1750

Park J., Obeysekera J., Barnes J. (2010a). Temporal Energy Partitions of Florida Extreme Sea Level Events as a function of Atlantic Multidecadal Oscillation. Ocean Science, 6, 587-593

Park J., Obeysekera J., Barnes J., Irizarry M., Park-Said W. (2010b) Climate Links and Variability of Extreme Sea Level Events at Key West, Pensacola, and Mayport Florida. ASCE Journal of Port, Coastal, Waterway and Ocean Engineering, 136 (6), 350-356

Park J., Obeysekera J., Irizarry M., Trimble P. (2011). Storm Surge Projections and Implications for Water Management in South Florida. Climatic Change, Special Issue: SLR in Florida: An Emerging Ecological and Social Crisis, Volume 107, Numbers 1-2, 109-128,

Parris, A. and L. Lacko 2009: Climate change adaptation in the San Francisco Bay: A case for managed realignment. Shore and Beach, vol. 77, pp 46-52.

Peltier WR (2004) Global glacial isostasy and the surface of the ice-age Earth: the ICE-5G (VM2) model and GRACE. Annu. Rev. Earth Planet Sci. 32:111-149

Pfeffer WT, Harper JT, O'Neel S (2008) Kinematic constraints on glacier contributions to 21st century sea-level rise. Science 321:1340–1343

Poag, C.W., Koeberl, Christian, Reimold, W.U., 2004, The Chesapeake Bay Crater, Springer, Berlin, Heidelberg, New York, 522 p.

Price SF, Payne AJ, Howat IM, Smith BE (2011) Committed sea-level rise for the next century from Greenland ice sheet dynamics during the past decade. Proc Natl Acad Sci USA 108:8978-8983

Rahmstorf, Stefan. 2007. A Semi-Empirical Approach to Projecting Future Sea-Level Rise. Science Vol. 315

Rahmstorf, S., Perrette, M., and M. Vermeer. 2011. Testing the robustness of semi-empirical sea level projections. Climate Dynamics. Volume 39, Numbers 3-4 (2012), 861-875

Rignot, E. J., G. Casassa, P. Gogineni, W. Krabill, A. Rivera, and R. Thomas. 2004. Accelerated ice discharge from the Antarctic Peninsula following the collapse of Larsen B ice shelf. Geophys. Res. Lett., 31, L18401, doi:10.1029/2004GL0697.

Rignot E, Bamber JL, van den Broeke MR, Davis C, Li Y, van de Berg VJ, van Meijgaard E (2008) Recent Antarctic ice mass loss from radar interferometry and regional climate modeling. Nat Geosci 1:106-110

Rignot E, Velicogna I, van den Broeke MR, Monaghan A, Lenaerts J (2011) Acceleration of the contribution of the Greenland and Antarctic ice sheets to SLR. Geophys Res Lett 38:L05503 doi:10.1029/2011GL046583

Rohling, E.J., Grant, K., Hemleben, C., Siddall, M., Hoogakker, B.A.A., Bolshaw, M. and Kucera, M. (2008) Letter. High rates of sea-level rise during the last interglacial period. Nature Geoscience, 1, 38-42. (doi:10.1038/ngeo.2007.28)

Sallenger, A., Doran, K, and Howd, P., 2012, Hotspot of accelerated sea-level rise on the Atlantic coast of North America, Nature Climate Change, DOI: 10.1038/NCLIMATE1597.

References

San Francisco Bay Conservation and Development Commission (SFBCDC) (2011) Living with a Rising Bay. Available: http://www.bcdc.ca.gov/BPA/LivingWithRisingBay.pdf. Accessed October 27, 2011.

Scambos, T.A., J.A. Bohlander, C.A. Shuman, and P. Skvarca. 2004. Glacier acceleration and thinning after ice shelf collapse in the Larsen B embayment, Antarctica. Geophys. Res. Lett., 31(18), L18402.

Schaeffer, M., Hare, W., Rahmstorf, S., and M. Vermeer. 2012. Long-term sea-level rise implied by 1.5 °C and 2 °C warming levels, Nature Climate Change, doi:10.1038/nclimate1584

Southeast Florida Regional Climate Change Compact Technical Ad hoc Work Group (SFRCCC). 2011. A Unified Sea Level Rise Projection for Southeast Florida. A document prepared for the Southeast Florida Regional Climate Change Compact Steering Committee. 27 p

Sella, G.F., Stein, S., Dixon, T.H., Craymer, M., James, T.S., Mazzotti, S. and R.K. Dokka. 2007. Observation of glacial isostatic adjustment in "stable" North America with GPS. Geophysical Research Letters, v. 34

Shinkle, K., and Dokka, R. K., 2004, Rates of vertical displacement at benchmarks in the lower Mississippi Valley and the northern Gulf Coast: NOAA Technical Report 50, 135 p.

Smits, A.J.M., Nienhuis, P.H., and Saeijs, H.L.F. 2006. Changing Estuaries, Changing Views. Hydrobiologia 565: 339–355.

Sun. H., Grandstaff. D., R. Shagam., Land subsidence due to groundwater withdrawal: potential damage of subsidence and sea level rise in southern New Jersey, USA, Environmental Geology, Vol.37, 1999, PP.290-296.

Thompson, P. T. (2011). Sea surface height: A versatile climate variable for investigations of decadal change. (Doctoral dissertation).

Timmermann, A., S. McGregor, and F.-F. Jin (2010) Wind effects on past and future regional sea level trends in the southern Indo-Pacific, J. Clim, 23, 4429-4437.

Titus, J.G., Hudgens, D.E., Trescott, D.L., Craghan, M., Nuckols, W.H., Hershner, C.H., Kassakian, J.M., Linn, C.J., Merritt, P.G., McCue, T.M., O'Connell, J.H., Tanski, J., and J Wang (2009) State and local governments plan for development of most land vulnerable to rising sea level along the US Atlantic coast. Environ. Res. Lett. 4

Titus James G. and Vijay Narayanan. (1995). The Probability of Sea Level Rise, U.S. Environmental Protection Agency. 186 pp. EPA 230-R95-008.

Tomasin A., and P. A. Pirazzoli. (2008). Extreme Sea Levels in the English Channel: Calibration of the Joint Probability Method, Journal of Coastal Research, 24(4C),1-13.

US Army Corps of Engineers (2011). Incorporating Sea-Level Change Considerations in Civil Works Programs, EC 1165-2-212.

Van den Broeke, M, Bamber, J, Lanaerts, J, and E. Rignot. 2011. Ice Sheets and Sea Level: Thinking Outside the Box. Surveys in Geophysics. DOI 10.1007/s10712-011-9137-Z

Vermeer M, Rahmstorf S (2009) Global sea level linked to global temperature. Proc Natl Acad Sci USA 106:21527–21532

Wang, K., R. Wells, S. Mazzotti, R. D. Hyndman, and T. Sagiya (2003), A revised dislocation model of interseismic deformation of the Cascadia subduction zone, J.Geophys. Res., 108(B1), 2026

Wang, K. (2007), Elastic and viscoelastic models of crustal deformation in subduction earthquake cycles, in The Seismogenic Zone of Subduction Thrust Faults, edited by T. Dixon and J. C. Moore, pp. 540–575, Columbia Univ. Press, New York.

Weeks, D., Malone, P., and L. Welling. 2011. Climate change scenario planning: A tool for managing parks into uncertain futures. Park Science, Volume 28, Number 1.

Yin J, Griffies SM, Stouffer RJ (2010) Spatial variability of SLR in twenty-first century projections. J Climate 23:4585–4607

Yin J, Overpeck JT, Griffies SM, Hu A, Russell JL, Stouffer RJ (2011) Different magnitudes of projected subsurface ocean warming around Greenland and Antarctica. Nat Geosci 4:524–528

Yin, J., Schlesinger, M., et al.. 2009, Model projections of rapid sea-level rise on the northeast coast of the United States, Nature Geoscience, 2(4), p. 262-266.

Yin, J. (2012) Century to multi-century sea level rise projections from CMIP5 models, Geophysical Res. Lett., 39.

Zhang K., Douglas B. C., Leatherman S. P. (2000), Twentieth-Century Storm Activity along the U.S. East Coast, J. Climate, 13, p1748-61.

Appendix

Sources for Appendix Definitions

NOAA Center for Operational Oceanographic Products and Services
http://tidesandcurrents.noaa.gov/est/faq.shtml#q1

American Meteorological Survey (AMS) Glossary of Meteorology
http://amsglossary.allenpress.com/glossary
Accessed October 27, 2011

National Geodetic Survey
http://www.ngs.noaa.gov/GEOID/geoid_def.html
Accessed October 27, 2011

NOAA AOML
http://www.aoml.noaa.gov/phod/amo_faq.php
Accessed April 27, 2012; NOAA

CPC
http://www.cpc.ncep.noaa.gov/products/analysis_monitoring/ensostuff/ensofaq.shtml#ENSO
Accessed April 27, 2012;

USGS Earthquake Hazards program
http://earthquake.usgs.gov/learn/glossary/?term=subduction
Accessed April 27, 2012

Glossary of Terms

Anomaly – A mean sea level anomaly occurs when the 5-month average of the interannual variation is greater than 0.1 meters (4 inches) or less than -0.1 meters.

Atlantic Meridional Overturning Circulation (AMOC) – The Meridional Overturning Circulation (MOC) is part of the global ocean circulation responsible for large-scale (on the order of 1000 km), low-frequency (interannual to multi-decadal), full-depth, meridional flux of mass, heat and freshwater. The Atlantic component of this circulation, the Atlantic Meridional Overturning Circulation (AMOC), has long been considered the dominant element of the MOC, in large part because the majority of water masses that compose the lower limb of the overturning circulation originate in the North Atlantic. (U.S. CLIVAR AMOC Planning Team, 2007: Implementation Strategy for a JSOST Near-Term Priority Assessing Meridional Overturning Circulation Variability: Implications for Rapid Climate Change. U.S. CLIVAR Report 2007-2, U.S. CLIVAR Office, Washington, DC, 20006, 23pp.)

Atlantic Multidecadal Oscillation (AMO) – The AMO is an ongoing series of long-duration changes in the sea surface temperature of the North Atlantic Ocean, with cool and warm phases that may last for 20-40 years at a time and with a difference of about 1°F between extremes. These changes are natural and have been occurring for at least the last 1,000 years.

El Niño Southern Oscillation (ENSO) – The ENSO cycle refers to the coherent, and sometimes very strong, year-to-year variations in sea-surface temperatures, convective rainfall, surface air pressure, and atmospheric circulation that occur across the equatorial Pacific Ocean. El Niño and La Niña represent opposite extremes in the ENSO cycle.

Eustatic sea level rise – Eustatic sea level rise is a change in global average sea level brought about by an increase in the volume of the world ocean (IPCC 2007b)

Geoid – The equipotential surface of the Earth's gravity field which best fits, in a least squares sense, global mean sea level.

Global mean sea level (GMSL) – Average height of the Earth's oceans. Global mean sea level can change globally due to (i) changes in the shape of the ocean basins, (ii) changes in the total mass of water (see eustatic sea level rise below) and (iii) changes in water density. Sea level changes induced by changes in water density are called steric. Density changes induced by temperature changes only are called thermosteric, while density changes induced by salinity changes are called halosteric (IPCC 2007b)

Joint Probability Method – Joint Probability Method refers to flood risk calculations using multiple variables, such as waves, sea level, river flow, and rainfall, to develop probabilities of flooding.

Local sea level (LSL) – The height of the water as measured along the coast relative to a specific point on land.

Mean Sea Level (MSL) – Refers to a tidal datum, or frame of vertical reference defined by a specific phase of the tide. Tidal datums are locally-derived based on observations at a tide station, and are typically computed over a 19-year period, known as the National Tidal Datum Epoch (NTDE). The present 19-year reference period used by NOAA is the 1983-2001 NTDE. Tidal datums must be updated at least every 20-25 years due to global sea level rise. Some stations are more frequently updated due to high relative sea level trends.

Pacific Decadal Oscillation (PDO) – The Pacific Decadal Oscillation (PDO) is the predominant source of inter-decadal climate variability in the Pacific Northwest (PNW). The PDO (like ENSO) is characterized by changes in sea surface temperature, sea level pressure, and wind patterns (Mantua 1997).

Relative sea level (RSL) – The height of the sea with respect to a specific point on land.

Steric (Halosteric & Thermosteric) – Sea level changes induced by changes in water density are called steric. Density changes induced by temperature changes only are called **thermosteric**, while density changes induced by salinity changes are called **halosteric**.

Subduction – Subduction is the process of the oceanic lithosphere colliding with and descending beneath the continental lithosphere.

Wind stress – The resistance per unit area caused by wind shear. For example, the wind stress on the sea surface applies a friction force that can drive ocean currents.

Wind stress curl – The vertical component of the (mathematical) curl of the surface wind stress. The large-scale, long-term averaged wind stress curl contains the principal information needed to calculate the wind-driven mass transport.

www.ingramcontent.com/pod-product-compliance
Lightning Source LLC
Chambersburg PA
CBHW081413170526
45166CB00010B/3326